Cambridge Primary

Hodder Cambridge Primary

Maths

Teacher's Pack

Foundation Stage

Ann and Paul Broadbent

HODDER
EDUCATION
AN HACHETTE UK COMPANY

Acknowlegement

We would like to thank the Department of Education for granting permission to reproduce the extracts on page 8, from page 12 of the Statutory framework for the Early Years Foundation Stage (EYFS), reference: DFE-00169-2017. Also refer to the link https://www.gov.uk/government/publications/early-years-foundation-stage-framework-2.

Every effort has been made to trace all copyright holders, but if any have been inadvertently overlooked, the Publishers will be pleased to make the necessary arrangements at the first opportunity.

Although every effort has been made to ensure that website addresses are correct at time of going to press, Hodder Education cannot be held responsible for the content of any website mentioned in this book. It is sometimes possible to find a relocated web page by typing in the address of the home page for a website in the URL window of your browser.

Hachette UK's policy is to use papers that are natural, renewable and recyclable products and made from wood grown in well-managed forests and other controlled sources. The logging and manufacturing processes are expected to conform to the environmental regulations of the country of origin.

Orders: please contact Hachette UK Distribution, Hely Hutchinson Centre, Milton Road, Didcot, Oxfordshire, OX11 7HH. Telephone: +44 (0)1235 827827. Email education@hachette.co.uk Lines are open from 9 a.m. to 5 p.m., Monday to Friday. You can also order through our website: www.hoddereducation.com

First published in 2018

This edition published in 2018 by Hodder Education,
An Hachette UK Company
Carmelite House
50 Victoria Embankment
London EC4Y 0DZ
www.hoddereducation.co.uk

Impression number   5 4

Year   2022

Cover illustration by Steve Evans

Illustrations by Jeanne du Plessis, Vian Oelofsen

Typeset in FS Albert 11 pt by Lizette Watkiss

Printed in the United Kingdom

A catalogue record for this title is available from the British Library.

ISBN 978 1 5104 3186 7

Contents

Term 2

Term 3

Photocopiable resources

Features of each unit

This Teacher's Pack must be used with the *Hodder Cambridge Primary Maths* Foundation Stage Activity Books and Story Books.

Icons indicate which components to use in each teaching sequence.

Resources are listed to help teachers prepare and plan for each practical activity.

Background information gives teachers important subject knowledge and pedagogical ideas on the content of each unit.

Learning objectives are provided for each unit to ensure progression and coverage of the Early Years Foundation Stage Early Learning Goals in Mathematics.

Key words are given to use, display and share with the learners in each lesson.

Activity Book teaching notes include useful teaching ideas to use with the content in the Activity Book.

Activity ideas give teachers a selection of practical activities to use with the learners as a class or in small groups.

Success criteria are based on the learning objectives to help identify what a learner should be able to do, know or understand by the end of the unit.

Assessment gives teachers further ideas for assessment, to use on completion of each Activity Book.

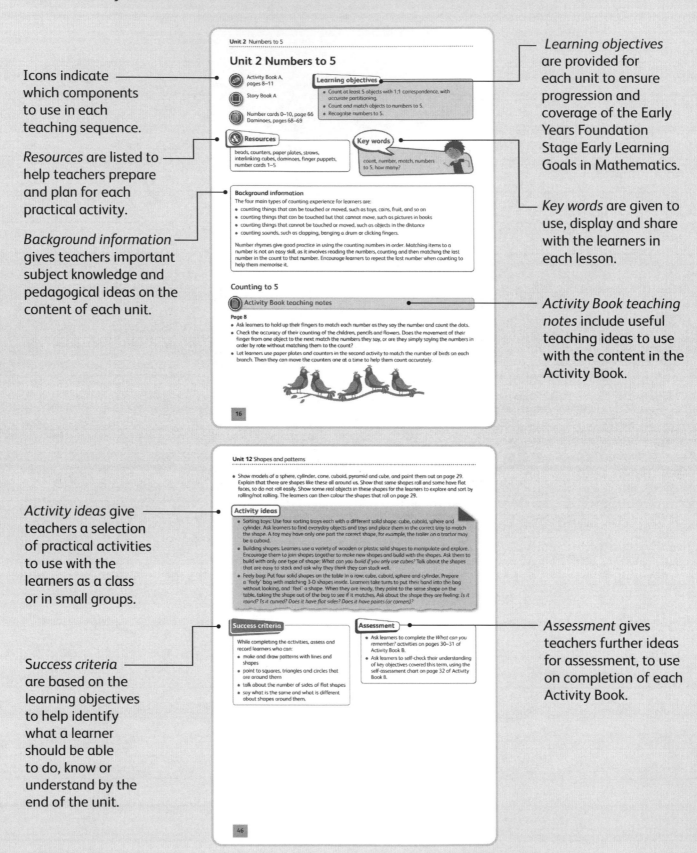

Introduction

About the series

Hodder Cambridge Primary Maths Foundation Stage is written by experienced authors and primary practitioners to reflect and prepare for the mastery approach in mathematics. The content and progression is based on coverage of the Early Years Foundation Stage Curriculum Framework and provides transition towards Stage 1 of the Cambridge Primary Mathematics Curriculum Framework and *Hodder Cambridge Primary Maths* Stage 1.

The Activity Books can be used as a stand-alone resource by teachers or parents/carers with one book per term. Or, the books can be used as part of a complete teaching programme in the classroom for ages four to five, with the content organised into 18 units. Teaching ideas and practical activities are provided in this Teacher's Pack for each of the 18 units.

An accompanying Story Book is also included for each term, containing a maths-based story to support the themes in the Activity Books. The Story Books can be used in class to explore the maths themes and can also be shared at home with parents or carers.

Components

Activity Books

There are three Activity Books (A, B and C) – one for each term.
Each Activity Book contains:

- six units, split into popular themes
- coverage of the objectives that underpin the Cambridge Primary Mathematics Stage 1 Curriculum Framework
- full coverage of the Early Years Foundation Stage Early Learning Goals in Mathematics
- motivating activities and practical tasks, following the mastery approach
- recap activities and a self-assessment chart at the end.

The units in the Activity Books consist of three main types of activity:

Explore: Some themes contain pictures, games, rhymes and diagrams for the teacher and learner to talk about and explore.

Learn: When helpful, a panel with a teaching point that clarifies the concept or skill is included, with a model or image to support understanding.

Practise: Practical and written activities, using resources, matching, colouring, drawing and writing, when appropriate, to complete the activities.

Teacher's Pack

The Teacher's Pack contains:

- a complete overview of the learning objectives for each term
- references to the Activity Book pages, the Story Books and the photocopy masters
- an overview of the learning objectives at the start of each unit
- teaching notes and practical activity ideas for each theme, built on a mastery approach, to embed mathematical understanding, skills and concepts
- background information on each theme, to support the subject knowledge of early years teachers
- key words for new vocabulary to use and display during the activities

- a suggested resource list for each unit
- success criteria at the end of each unit to assist formative and summative assessment
- photocopy masters to support the activities.

 Story Books

There are three Story Books (A, B and C) to support the maths concepts in the Activity Books. The Story Books can be read as a class, or in groups, or can be sent home for learners to share with their parents or carers.

Each Story Book contains:

- colourful illustrations and a lively story covering the maths themes in the Activity Books
- teacher/parent/carer notes on each page, giving opportunities to talk about the maths in the picture or story
- additional activities at the end of the book for the learner to complete.

Structure, scope and sequence

The structure and content of each unit is based on the Early Years Foundation Stage Curriculum Framework, and provides transition towards Stage 1 of the Cambridge Primary Mathematics Curriculum Framework. A spiral curriculum model is used to make sure there is both curriculum coverage and careful progression of concepts and skills over the stage.

The stage is divided into three terms, with each term containing 6 units (18 units in total). Each unit is intended to last approximately two weeks.

Early Years Foundation Stage Curriculum Framework

The Early Years Foundation Stage Curriculum Framework sets out the end-of-year expectations for ages four to five *(Foundation Stage 2/Reception/Kindergarten Upper)*, as defined by these Early Learning Goals in mathematics:

> *Mathematics*
>
> This involves providing children with opportunities to:
> - *practise and improve their skills in counting numbers, calculating simple addition and subtraction problems*
> - *describe shapes, spaces and measures.*
>
> **Numbers:** *Children count reliably with numbers from 1 to 20, place them in order and say which number is one more or one less than a given number. Using quantities and objects, they add and subtract two single-digit numbers and count on or back to find the answer. They solve problems, including doubling, halving and sharing.*
>
> **Shape, space and measures:** *Children use everyday language to talk about size, weight, capacity, position, distance, time and money to compare quantities and objects and to solve problems. They recognise, create and describe patterns. They explore characteristics of everyday objects and shapes and use mathematical language to describe them.*
>
> The extract is from: Statutory framework for the Early Years Foundation Stage, Department for Education.

Hodder Cambridge Primary Maths Foundation Stage has full coverage of these Early Learning Goals and careful 'small-step' progression built in to support learners' understanding of skills and concepts.

A mastery approach

The *Hodder Cambridge Primary Maths* series adopts a mastery approach, which involves helping learners develop a deep understanding of mathematical ideas and concepts, so that they can make connections between different areas of mathematics. While fluency with procedures such as calculation skills is important, learners who have mastered a particular aspect of mathematics fully are able to apply what they have learnt to solve unfamiliar problems.

Hodder Cambridge Primary Maths Foundation Stage uses elements of a mastery approach that are relevant for early year's learners:

- Sufficient time is allowed on each theme for depth of coverage and practice, with six clear units per term.
- There is careful 'small-step' progression of each unit, with a spiral approach so that themes are revisited and developed each term.
- Learning objectives and expected outcomes are provided so it is clear which concepts and skills need teaching and learning.
- Bruner's Concrete-Pictorial-Abstract (CPA) model underpins each unit with appropriate concrete experiences, pictures and language, helping learners make sense of the mathematics.

Representation

Concrete-Pictorial-Abstract (CPA)

Hodder Cambridge Primary Maths is based on the belief that mathematical understanding is developed through using concrete, pictorial and abstract (or symbolic) representations. Practical resources and diagrammatic representations are used throughout for teachers and learners to use and make sense of a concept or problem. Learners will need to explore the various representations and be allowed the opportunity to choose which representations they use for a particular activity.

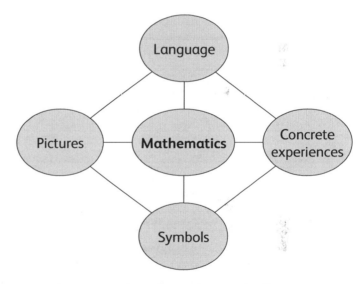

Mathematical language

Language is another important element of representation, with the connections between language, concrete experiences, pictures and symbols helping children make sense of the mathematics. This means that discussions and talk using accurate mathematical language is very important in the classroom for sharing ideas, addressing misconceptions and developing reasoning skills. Key words are given in each unit. Teachers need to plan the introduction of this new vocabulary into the activities, and provide opportunities for learners to rehearse and use the words on a regular basis.

Assessment

Formative assessment

Ongoing, formative assessment is central to teaching and learning for mastery, particularly in ensuring that learners have achieved more than a superficial understanding, and to avoid moving on too quickly.

Formative assessment should be used to inform the next steps in learning, and may influence changes in planning and therefore the next lessons. Formative assessment is a cycle, finding out what learners know, moving learning forward, finding out how that learning has changed (what they know now) and planning the next steps. Where you find that learners are still unsure, stop and take time to revisit an idea or skill, change the activity or context and then move onto new learning when learning is secure. Assessment is about you (and the learners) continually reflecting on learning, and ensuring that teaching is in line with learning.

Introduction

Throughout each unit, there are continual opportunities for assessment. As the teacher, you will probe conceptual and procedural understanding through questioning and observation as you model and teach. The way learners respond to the modelling and teaching provides you with valuable information on what to spend a little more time and what to move through quickly, as well as information on individual needs.

Learning objectives and success criteria

A learning objectives overview is provided at the beginning of each term in this Teacher's Pack. Suggestions for success criteria are also given at the end of a unit. The success criteria are used to help assess the outcome of the learning that has taken place. They are, in effect, what the successful learning will 'look' like once the learning objectives have been met. At the end of each lesson, ask the learners to reflect on what they have learnt, and check their understanding against the success criteria. The self-assessment chart at the end of each Activity Book includes the key objectives, and are designed to give learners the opportunity to demonstrate what they know and the concepts they have mastered.

Practical resources

Visual representations, manipulatives and concrete resources are extremely important in helping learners to develop a conceptual understanding of what they are learning.

The following key manipulatives are included in teaching activities and the Activity Books:

- Straws
- Digit cards
- Counting sticks
- Coloured counters
- 3-D shapes
- 1–6 spinners

- Number rods
- Coins
- Interlocking cubes
- 2-D shapes
- Clock faces
- Number cards.

The following key visual representations are included in the Activity Books:

- Number tracks
- Number lines

- Ten frames
- The bar model.

There are other resources in an early years' classroom that can be used to enrich mathematics teaching and learning. The early years' environment often includes specific settings for learners to experience. Opportunities can be taken to use these settings to develop mathematical concepts and skills. These settings include sand (dry and wet), water, writing or mark-making, reading or books, small-world toys, outdoors, role-play, construction, music, 'home' corner and Information Technology (IT).

1	2	3	4	5	6	7	8	9	10
11	12	13	14	15	16	17	18	19	20
21	22	23	24	25	26	27	28	29	30
31	32	33	34	35	36	37	38	39	40
41	42	43	44	45	46	47	48	49	50
51	52	53	54	55	56	57	58	59	60
61	62	63	64	65	66	67	68	69	70
71	72	73	74	75	76	77	78	79	80
81	82	83	84	85	86	87	88	89	90
91	92	93	94	95	96	97	98	99	100

0 1 2 3 4 5 6 7 8 9 10 11 12 13 14 15 16 17 18 19 20

Learning objectives overview for Term 1

Units	Themes	Learning objectives	Activity Book pages	Preparation for Cambridge Primary Maths Stage 1 (Framework Code)
Unit 1 Sorting and counting	**Sorting**	Sort and match objects in a set.	4–5	1Dh1, 1Nn3
	Matching	Match items that go together.	6	
	Counting how many	Count reliably at least 5 objects, recognising that when rearranged, the number of objects stays the same. Understand that the last number in the count represents the set as a whole.	7	
Unit 2 Numbers to 5	**Counting to 5**	Count at least 5 objects with 1:1 correspondence, with accurate partitioning.	8	1Nn1, 1Nn2, 1Nn3
	Matching numbers	Count and match objects to numbers to 5.	9	
	Counting and numbers to 5	Recognise numbers to 5.	10–11	
Unit 3 Counting forwards and backwards	**Counting on and back**	Count forwards and backwards to 5.	12–13	1Nn1, 1Nn2, 1Nn3
	Zero to five	Recognise zero as the empty set. Recognise numbers to 5 and beyond in the environment.	14–15	
	Writing numbers	Count in order and match numbers to objects. Write numbers to 5.	16–17	
Unit 4 Addition and subtraction	**One more and one less**	Add one more to a set of objects to 5 and say how many. Take one away from a set of objects to 5 and say how many.	18–19	1Nn8, 1Nc12, 1Nc8, 1Nc9
	Putting together	Combine and count all the objects in two sets to make a total up to 5.	20	
	Making totals	Partition numbers to 5 in different ways.	21	

Units	Themes	Learning objectives	Activity Book pages	Preparation for Cambridge Primary Maths Stage 1 (Framework Code)
Unit 5 Measures and time	Comparing objects	Use everyday language to talk about the size of objects.	22	1MI1, 1MI2, 1MI3, 1Mt1, 1Mt3
	Comparing length	Compare two items by length or height and say which is longer and which is shorter.	23	
	Long and short	Compare two items by length or height and say which is longer and which is shorter.	24	
	Day and night	Recognise things that happen in the morning, afternoon and night. Recognise and name days of the week, using them in context.	25	
Unit 6 Shapes and patterns	I spy ... shapes	Describe and recognise simple flat shapes.	26–27	1Gs1, 1Gs2
	Patterns	Draw and describe simple line patterns.	28	
	Solid shapes	Build and describe models made with boxes and objects.	29	

Unit 1 Sorting and counting

Activity Book A,
pages 4–7

Story Book A

Learning objectives

- Sort and match objects in a set.
- Match items that go together.
- Count reliably at least five objects, recognising that when rearranged, the number of objects stays the same.
- Understand that the last number in the count represents the set as a whole.

Resources

sorting trays, different types of cutlery, beads, counters, cubes, shape tiles, ribbons of different and same lengths, small balls, quoits, beanbags, old greeting and birthday cards, old catalogues, plastic jars and containers and lids, coloured pencil crayons

Key words

sort, count, group, numbers 1–5, set, compare, same, different

Background information

Sorting objects supports much of the early conceptual understanding of similarities and differences between items, numbers and shapes. The question, *What is the same and what is different?,* is an excellent one for giving learners the chance to identify the features and properties of objects, shapes and numbers. It is also an opportunity to compare objects before formal measuring.

Sorting

Activity Book teaching notes

Pages 4–5

- Encourage learners to choose their own way of sorting the clothing. The items could be sorted by design, for example all the spotty items in the spotty box, or by use, for example all the T-shirts in the T-shirt box.

- Talk about the shapes of the keys and what makes one key the same or different to another key, before learners colour the keys on page 5.

- Allow learners to do some practical sorting of cutlery before looking at the second activity on page 5. Learners could sort by type of cutlery (for example, all the spoons or all the forks) or by sets of cutlery (matching knife, fork and spoon).

Activity ideas

- Tidy away: Encourage learners to tidy up resources, as this activity provides good opportunities for sorting. Ask learners to sort their toys or equipment in different ways. They could sort by colour, type, size and shape, or according to the materials from which the items are made.

- Sorting stuff: Keep a collection of classroom 'junk' and discuss different ways of sorting it with learners. For example, sort all the red items, items used for drawing, items for eating and drinking, and so on. Ask: *What is the same about these items? Find two items that are the same.*

- Odd-one-out: Take some items that are the same, perhaps all red items or all coins, from a sorting tray. Add something that does not belong to the set. Ask learners to find the odd-one-out. This may be repeated with learners taking it in turns to make a set with an odd item.

- Sorting pictures: Use old catalogues, greeting cards or birthday cards for this activity. Ask learners to cut out the pictures. They can choose their own criteria for sorting, for example sets of red shoes, winter coats, toy cars, tricycles, and so on.

- Sorting equipment: Make a collection of small balls, quoits and beanbags. Discuss with the learners how to sort them, for example by use or by colour. Let learners take it in turns to find an item that belongs to the set of:
  - things you can throw
  - things that will roll
  - blue things.

Extend the activity to sorting for two criteria, for example: *It is red and it has holes.*

Matching

 Activity Book teaching notes

Page 6

- Talk about the different socks on the washing line, using language such as *long*, *short*, *striped* and *spotted*. Learners will need six different coloured pencils to colour the pairs of socks.

- Model the method of joining pairs of shoes with a line in the second activity, if necessary. If you prefer, you could also place coloured counters on the shoes to show each pair.

Activity ideas

- Match jars and lids: Put some jars and their lids on a tray. Ask learners to match the jars and lids. As each lid is chosen, ask whether it is the right size, or too small or too big. Repeat with other containers and their lids. Fill one of the containers with sand, water or dried peas. Talk about the concepts of 'empty', 'full' and 'nearly full'. Discuss which container holds the most. Check learners' predictions by pouring from one container to the other.

- Match ribbons: Use a selection of different ribbons, with some of the same length and others of the same colour. Ask learners to choose two ribbons that match. Ask them how the ribbons match and if there are any others that match in length or colour. If learners find two ribbons that don't match in length, ask: *Which is the longer ribbon? Which is the shorter ribbon?* Encourage them to lay the ribbons side by side, with one pair of ends aligned, so that they can make a direct comparison.

Counting how many

 Activity Book teaching notes

Page 7

- In the first activity, the animals are spread out randomly, which makes them more difficult for learners to count. Demonstrate counting incorrectly, for example by touching animals twice or missing some out, and talk about what went wrong with the count. Ask learners to model counting carefully, possibly touching each animal or covering them with a counter so that no animals are missed.

- In the second activity, ask the learners how the flowers are the same and how they are different. Count the petals on one of the flowers and then ask learners to find another flower with the same number of petals. Let them shade these flowers in the same colour.

Activity ideas

- Counting movable objects: Use collections of objects, such as cubes, counters or toy cars, for this activity. Ask learners to take a few objects and count them. Check the counting technique of learners who appear to miscount. Do they move the items that have been counted away from those waiting to be counted? Do they match the action of touching to the number word? Do they use the correct sequence of counting words? Encourage confident counters to take a handful, or even two handfuls, and to count how many objects they have altogether.

- Quick recognition of small quantities between 0 and 4: Show learners small sets of counters in your hand. Keep the sets to between 0 and 4. Ask learners to say how many counters there are. Aim for immediate responses, so the learners are subitising (knowing how many objects there are without counting).

- Counters game: Let learners work in pairs, with each learner having five counters. They take turns to show their partner some or all of their five counters. How quickly can their partner say how many counters there are?

Success criteria

While completing the activities, assess and record learners who can:
- join matching pairs
- count objects in a set by moving them one at a time
- count how many cubes there are in a set.

Unit 2 Numbers to 5

 Activity Book A, pages 8–11

 Story Book A

 Number cards 0–10, page 66
Dominoes, pages 68–69

Learning objectives

- Count at least 5 objects with 1:1 correspondence, with accurate partitioning.
- Count and match objects to numbers to 5.
- Recognise numbers to 5.

 Resources

beads, counters, paper plates, straws, interlinking cubes, dominoes, finger puppets, number cards 1–5

 **Key words**

count, number, match, numbers to 5, how many?

Background information

The four main types of counting experience for learners are:

- counting things that can be touched or moved, such as toys, coins, fruit, and so on
- counting things that can be touched but that cannot move, such as pictures in books
- counting things that cannot be touched or moved, such as objects in the distance
- counting sounds, such as clapping, banging a drum or clicking fingers.

Number rhymes give good practice in using the counting numbers in order. Matching items to a number is not an easy skill, as it involves reading the numbers, counting and then matching the last number in the count to that number. Encourage learners to repeat the last number when counting to help them memorise it.

Counting to 5

 Activity Book teaching notes

Page 8

- Ask learners to hold up their fingers to match each number as they say the number and count the dots.
- Check the accuracy of their counting of the children, pencils and flowers. Does the movement of their finger from one object to the next match the numbers they say, or are they simply saying the numbers in order by rote without matching them to the count?
- Let learners use paper plates and counters in the second activity to match the number of birds on each branch. Then they can move the counters one at a time to help them count accurately.

Activity ideas

- Count to check: Ask the learners to use counting materials, such as counters and cubes. Ask them to take a handful and to guess how many they have, then count to check. Ask learners for ideas on how to record what has been done. These ideas could include drawing or colouring in squares on strips of squared paper.
- Dominoes: Dots on dominoes are very useful for counting. Ask learners to find different dominoes with a total of four dots, for example. Encourage learners to look at the pattern of the dots when they count. Cut out and use PCM 3: Dominoes on pages 68–69 of this Teacher's Pack.
- Finger puppets: Get a set of finger puppets and act out stories with different numbers of puppets being held up for the learners to count.

Matching numbers

 Activity Book teaching notes

Page 9

- Count the carriages on the train and emphasise the last number in the count as being the total number of carriages. This is the *Cardinal Principle*, which means the last number word of an array of counted items represents the set as a whole.
- The second activity is an open task, with the learners choosing how many cars to colour. Then they can trace the number to match.

Activity ideas

- Show me: Each learner has a set of number cards from 1–5. Use PCM 1: Number cards on page 66 of this Teacher's Pack. Play 'Show me' activities where each learner shows a number card, by holding it up in the air. Hold up a large number card and say: *Show me a number like this. What does it say?* Ask learners to hold up selected numbers: *Show me number 4. Show me number 2.*
- Number match: Play 'Show me' activities as above, but this time ask learners to match their number to a cube tower that you hold up. Count out the cubes together and then let learners choose a number card to show.

Counting and numbers to 5

 Activity Book teaching notes

Pages 10–11

- Check that learners can recognise that the last number in the count is a label that shows how many items there are. Encourage them to count by touching each item as they count. They can make their own brick towers using interlocking cubes by standing them together in line in the correct order.
- Remind learners that they need to count carefully, to make sure they do not count the same item twice or miss out any items. Relate the last number in the count to the total.
- Encourage the learners to say the numbers aloud or silently as they practise writing them. Model the start position for each number, with the numbers 4 and 5 needing two strokes to complete the numbers.

Activity ideas

- Counting objects: Use collections of objects, such as cubes, straws and counters, as well as a set of number cards from 1–5. Ask learners to take a handful of objects and count them, and then to match the objects to a number card. Check the counting technique of learners who appear to miscount.

- Behind the wall: Play 'Behind the wall' activities by sliding a number card up from behind a 'wall'. A 'wall' can be a piece of card, a book or any other suitable screen. Have part of a number just peeping over the top of the wall. Ask: *Show me which number this might be. Could it be any other number? Show me.* Keep showing a little more of the number and repeating the instruction.

- Number match: Learners work in pairs, with each learner having five cubes and a set of number cards from 1–5. They take turns to show their partner some or all of their five cubes. How quickly can their partner show how many there are with a matching number card?

Success criteria

While completing the activities, assess and record learners who can:

- count objects and sounds
- count and match objects to a number card
- read the numbers to 5.

Unit 3 Counting forwards and backwards

 Activity Book A, pages 12–17

 Story Book A

 Five frames, page 74
Number cards 0–10, page 66

Learning objectives

- Count forwards and backwards to 5.
- Recognise zero as the empty set.
- Recognise numbers to 5 and beyond in the environment.
- Count in order and match numbers to objects.
- Write numbers to 5.

 Resources

beads, buttons, plastic pots, straws, counters, interlocking cubes, sand and water trays, bucket, five frames, number cards to 5

Key words

count, match, numbers to 5, zero, nothing, empty set, compare, order, forwards, backwards

Background information

Use counting rhymes and rhythms to give learners experience of counting numbers to 5 and back. To begin with, count forwards and backwards to 3, then to 5 and on to 10. With good counting skills, learners can count 'how many' accurately in a set. They should also develop the ability to subitise, which involves judging how many there are in a small set without counting. They may be able to subitise up to three or four items.

Let learners trace numbers in the air and in sand, as well as practise writing each number. Check learners start at the correct position and relate each written number to the last number in the count.

Counting on and back

 Activity Book teaching notes

Pages 12–13

- Use real objects (such as cubes) to represent the frogs while reading or singing the rhyme. Put the 'frogs' on a table and a bucket on a chair so that they physically move into a 'pool' during each verse. Repeat this activity in the water tray area, if possible.

- Check that learners count the objects in the picture accurately. They may lose their place or miss out objects, so discuss strategies such as touching each object or placing counters on them to help count.

Activity ideas

- Count and hop: Ask learners to pretend to be frogs and to hop forward while counting aloud slowly. They could change direction at the end of the count sequence, for example:

  One   two   three   four   five
  hop   hop   hop   hop   hop (change direction)

- Dropping cubes: Drop five interlocking cubes one by one into a tray (or other suitable resource), so each one makes a sound as the learners say the numbers in sequence. Ask five learners to hold a number card from 1 to 5 so they are standing in the correct order. As the cubes are dropped, the learners should hold the card up to match the number.

- Boston wave: Let learners sit in a circle and count slowly together: *One-two-three-four-five, one-two-three-four-five, one …* Decide which way round the circle the 'Boston wave' goes. As 'one' is chanted, point to a learner who quickly stands up, then immediately sits down. Point to the next learner in the circle as 'two' is chanted, and he or she then also stands up and sits down quickly. Repeat the 'stand–sit' action going around the circle. No more than one learner should be standing at any one time and the 'stand–sit' action should be done in time to the count rhythm. With practice, and as learners pick up the beat, there will be no need to point; learners will automatically continue the 'Boston wave' round the circle.

Zero to five

 Activity Book teaching notes

Pages 14–15

- Model the first activity. Show an empty bowl. Say there is nothing in there and show this is written as 0 (say *zero*). Show an open hand and count down from 5 to 0, folding in the fingers until you just show a fist – emphasise that this shows zero fingers.

- Zero can be understood as an empty set or the position on a number line before 1. Make sure learners experience both ways of understanding zero.

- Use five frames and counters to represent the numbers to 5 (and then to 10). Place counters in the frames and count them. Number frames are a very good model for learners to use to help gain number sense. Also use PCM 9: Five frames on page 74 of this Teacher's Pack for additional practice.

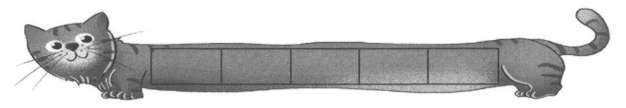

Activity ideas

- Shake and rattle: Put different numbers of beads or buttons into a set of five or six small plastic pots. Make sure two of the pots are empty. Mix the pots up and ask learners to guess which pots have zero beads in them. They check if they have guessed correctly by picking up the pot and shaking it to see if it rattles.

- Show me: Each learner has a set of number cards from 0–5. Use PCM 1: Number cards on page 66 of this Teacher's Pack. Play 'Show me' activities where each learner shows a number card by holding it up in the air. Hold up a large number card and ask: *Show me a number like this. What does it say?* Ask learners to hold up selected numbers: *Show me number 3. Show me zero.*

- Number match: Play the 'Show me' activities as above with learners, but this time, let them match their number to a cube tower that you hold up. Count out the cubes together and then they choose a number card to show. Make sure zero is included.

Writing numbers

Activity Book teaching notes

Pages 16–17

- Encourage learners to count numbers of objects aloud, as this helps them to recognise and write numbers. In the activity on page 16, demonstrate how to write the final number in the count.

- Check that learners colour the circles accurately to match each number. When they have finished, cover a number with a counter or finger and ask learners to say what they think is the missing number. They should count the circles to check before revealing the number.

- Encourage learners to say the numbers aloud or silently in their head as they practise writing the numbers on both activity pages. Model the start position for each number, with the numbers 4 and 5 needing two strokes to complete the numbers.

Activity ideas

- Numbers in the air: For numbers 4 and 5, use two colours (red and blue) to write the two stages (labelled (1) and (2) on the activity pages) of each number on large pieces of card. Ask the learners to stand and point to each number, tracing in the air first the red part (1), and then the blue part (2). This process will help learners to see the different movements needed when writing these numbers.

- Behind the wall: Play the 'Behind the wall' activity by sliding a number card up from behind a 'wall'. A wall can be a piece of card, a book or any other suitable screen. Have part of a number just peeping over the top of the wall. Ask: *Show me which number this might be. Could it be any other number? Show me.* Keep showing a little more of the number and repeating the instruction.

Success criteria

While completing the activities, assess and record learners who can:
- count to 5 and then back again to zero
- know that 0 means zero or 'nothing'
- read numbers around them
- count using a five frame
- write the numbers to 5.

Unit 4 Addition and subtraction

 Activity Book A,
pages 18–21

 Story Book A

 Partition circles, page 76

 Learning objectives

- Add one more to a set of objects to 5 and say how many.
- Take one away from a set of objects to 5 and say how many.
- Combine and count all the objects in two sets to make a total up to 5.
- Partition numbers to 5 in different ways.

 Resources

beads, counters, interlocking cubes, toy animals, coins number track 0–10, number rods, floor number track to 10, string

Key words

add, take away, combine, count, makes, total, more, less, how many? partition, group, break up

Background information

Once confident at counting to 5 and then to 10, recognising one more than a number is a first step towards addition. It reinforces and moves learners towards an understanding of ordinal numbers and positional numbers.

Give learners the opportunity to combine sets of objects to make totals to 5. Introduce the language of addition and link it to counting to make totals.

Subtraction is introduced as 'taking away'. Explain to learners that when you take away an object from a group of objects, there is now one less than the original number.

One more and one less

Activity Book teaching notes

Pages 18–19

- The 'Bus stop' game gives learners practice in adding one and taking away one as they move from one bus stop to the next. The 'go to the stop that is 1 more or 1 less' depends on the flip of their coin. The coin needs preparing beforehand, with stickers on either side for the instructions '1 more' and '1 less'.
- Make sure learners understand that if they land on a bus stop with a passenger waiting, they can add 1 more and move their counter. The number track on page 19 can be used to match the numbers learners land on, so they can see one more and one less more clearly.

Activity ideas

- Finger flash: Ask learners to hold up a number of fingers in response to 'and 1 more' or 'and 1 less' questions, for example: *Show me three fingers – and one more. Show me five fingers – and one less.*

- More and less: Provide learners with trays of small countable items, such as interlocking cubes, counters and toy animals. Ask: *Put four counters together – now add one more. Put three cubes in a row – now show me one less.*

- Floor number track 0–10: Use painted numbers on the playground or numbers written on card and placed in sequence on the ground. Ask learners to hop forwards and backwards along the track one number at a time. When they hop backwards, it is important that they move back without turning around. The idea of simple inverses can be developed. Ask: *Stand on 4. Make one hop forward. Which number are you standing on now? Make one hop back. Which number are you standing on now?*

Putting together

 Activity Book teaching notes

Page 20

- This page gives learners the opportunity to practise combining sets of objects and then recording them. Reinforce the language of 'putting together', 'adding' and 'finding totals'.

- Use small sets of countable objects to combine, and then observe the learners in the methods they use to work out the totals. Learners are likely to count all the objects when adding them together. Encourage them to count on from the first total – a common error is to count the second set from 1 again. A strategy that learners will use later is to count on from the largest number.

Activity ideas

- Combining: Provide learners with counters or cubes. Ask: *Show me three red cubes. Show me two green cubes. How many altogether?* Repeat for other quantities and use appropriate language, such as: *Add them together. What is the total?*

- Combining machine: Let learners make a combining machine in pairs, with two funnels or tubes leading into a box. They can show two sets of cubes, such as two and three, going into the machine. Ask them to say how many will come out of the machine and then to check this by counting. Repeat for other amounts.

Making totals

 Activity Book teaching notes

Page 21

- Partitioning numbers in different ways is important for mental fluency and for learners to get a 'feel' for numbers, particularly with numbers to 5 and then to 10.

- Use cubes, counters or number rods to represent a starting number and then ask learners to show different ways of partitioning or breaking up that number. Recording diagrams like those on page 21 can be used to show the different ways that numbers can be partitioned.

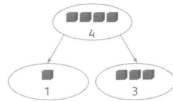

Activity ideas

- In the one hand: Each learner has five small objects, such as counters or coins. They place some counters in one hand and the rest in the other hand, and then they close their hands. (They may put all the counters in one hand.) In turn, each learner opens one hand saying 'three and … (pausing before opening the other hand) two'. Learners will then hear the number combinations making five over and over again. Eventually, some learners will be able to say how many objects are in their second hand before they open it. Repeat this for different totals to 5.

- The story of 5: Put a large number 5 in the centre of a poster and ask learners to find different ways to make the total 5. Objects can be stuck on the poster or drawn to represent the different combinations.

- Partitioning a set: Give each learner a set of small objects on their workspace and a length of string with which to partition the set. They begin by counting the total number in their set (the number of objects in the sets may be the same or different for all the learners). Each learner divides their set into two, using the string. Then they take turns to say how they divided or partitioned their set. For example: *I have put my four into a two and a two. Five can be shared into two and three.* After each partition, they combine the sets together again and repeat with different partitions. Some learners may choose to include an empty set in their partitioning. You could use PCM 11: Partition circles (instead of string) on page 76 of this Teacher's Pack.

Success criteria

While completing the activities, assess and record learners who can:
- add one more to a set and say how many
- take one away from a set and say what is left
- put two groups of objects together and count the total
- break up 5 cubes in different ways and show 4 and 1, then 3 and 2.

Unit 5 Measures and time

 Activity Book A, pages 22–25

 Story Book A

Learning objectives

- Compare two items by length or height and say which is longer and which is shorter.
- Use everyday language to talk about the size of objects.
- Recognise things that happen in the morning, afternoon and night.
- Recognise and name days of the week, using them in context.

 Resources

cubes, small objects for comparing size and length (leaves, pencils, coins, shells), ribbons, string, modelling clay, food tins, small-world toys, shoes, sugar paper, old magazines

Key words

length, size, small, big, large, smaller, larger, bigger, long, longer, short, shorter, longest, shortest, tall, taller, tallest, time, day, night, evening, afternoon, morning

Background information

Use appropriate vocabulary when comparing the size and length of objects, so that learners become familiar with the language of measurement. Comparative language and opposite terms (for example, tall and short) are key ideas as preparation for ordering, and then measuring length and size. Use language related to size: *big, small, little, thin, wide, huge* and *tiny*. Use these key words for comparing: *longer, shorter, taller* and *wider*.

Talk about the morning, afternoon and night as an everyday occurrence in the classroom so the learners start to understand the rhythm and timings in a day. The days of the week can be introduced in a similar way.

Comparing objects

 Activity Book teaching notes

Page 22

Give learners sets of different objects to compare, such as leaves, pencils, coins and shells, before they compare the size and height of the pictures of objects on page 22. Ask them to look for similarities and differences. Then use the language of comparison to describe and compare the objects, and to identify a single item. Ask: *Which is the largest? Longest? Smallest? Shortest?*

Activity ideas

- Food tins: Compare the sizes of different tins of food. Ask: *Which is the biggest? Which is the smallest? Which is the tallest? Which is the heaviest? Put the tins in order of size.*
- Small-worlds: Use small-worlds such as toy farmyards, houses or garages, and line up animals, people or cars so that they can be compared. Ask learners to compare their sizes and to find the smallest, largest, tallest, longest and shortest items.
- Comparing shoes: Put a collection of shoes in a row. Ask learners: *Which is the biggest? Smallest? Heaviest?* Pick up two shoes that do not match. Ask: *Which is the longer shoe?*

Comparing length

 Activity Book teaching notes

Page 23

This page gives the learners the opportunity to compare lengths of objects and to identify the longest and the shortest in a set. Measuring is not being taught at this stage. Hold up four lengths of string and ask learners how they know which one is the longest. Ask what they could do to make sure they are correct, and then demonstrate lining them up next to each other to see which is the longest. Then they can use this method on page 23 by using lengths of string for each fishing rod.

Activity ideas

- Collage: Use ribbons, string and yarn, letting learners add them to paintings or collage pictures. Learners can make:
  - a cat with long whiskers
  - a mouse with a very long tail
  - a child with long hair.
- Modelling clay models: Ask learners to make and describe modelling clay models using language of size and length. For example:
  - a very *long* worm
  - a frog with *short* legs
  - a *large* fish and a *smaller* fish.

Long and short

Activity Book teaching notes

Pages 24

Distinguish the '-er' comparison language, usually used when comparing two objects, from '-est' when a set of objects is compared and a single one is selected: Ask: *Which is the longest? Shortest? Largest? Smallest?* Use plenty of practical comparisons to make this clear to learners.

Activity ideas

- Ribbon match: Show the learners two ribbons and ask: *Which is the longer ribbon? Which is the shorter ribbon?* Ask learners to take it in turns to choose two ribbons and to show which is longer and which is shorter. Encourage them to lay the ribbons side by side, with one pair of ends aligned, so that a direct comparison is made.

Day and night

Activity Book teaching notes

Page 25

- Talk about the morning, afternoon and night when opportunities arise in the classroom, so the learners get a feel for the rhythm and timings in a day. The activity page concentrates on just one day and night, so the focus can be on the features, similarities and differences of these two times.
- The second activity introduces the days of the week. This activity needs to be used each day in the classroom to reinforce the order, names and cycle in each week.

Activity ideas

- The school day: Make a chart for the school day and let learners help to complete it, showing what will happen during the day. Encourage the use of language such as *before*, *next* and *after*. At the end of the day, encourage learners to remember what has happened during the day, in sequence, and compare this with the chart.
- The week: Make a chart for the week, showing the main events that will occur. This can be checked at the end of the day. Discuss days of the week and their order.
- Day and night: Prepare two large pieces of sugar paper and place them side by side to show day and night. Ask learners to draw or find day and night pictures from magazines. They should then stick daytime pictures on one half of the paper and night-time pictures on the other half. Display in the classroom.
- Tell me about …: Talk about the sequence of a day: morning, afternoon, evening and night. Discuss daily events such as breakfast, going to school, lunch, dinner and bedtime.

Success criteria

While completing the activities, assess and record learners who can:

- compare the length of two objects and say which is longer
- compare objects and say which are bigger or smaller
- talk about things that happen in the morning, afternoon, evening and night
- recall the days of the week.

Unit 6 Shapes and patterns

 Activity Book A, pages 26–29

 Story Book A

Learning objectives

- Describe and recognise simple flat shapes.
- Draw and describe simple line patterns.
- Build and describe models made with boxes and objects.

Resources

2-D shape tiles, models of 3-D solid shapes, patterned fabrics, wallpaper, trays of beads and laces, items for printing (cotton reels, sponge off-cuts, feathers, toothbrushes), cubes, pegboards, junk materials for model dragon, building bricks

Key words

flat shapes, solid shapes, circle, triangle, square, rectangle, cube, cuboid, cone, cylinder, make, build, draw, curved, straight, hollow, solid, flat, side, corner, point, pattern, line

Background information

The two basic ideas for learners to understand about shapes involve classifying shapes and transforming them. Classifying involves putting shapes into the same sets according to attributes that are 'the same'. Then we can change shapes in different ways and look at what is different. Encourage learners to always look at the similarities and differences between shapes. Ask key questions, for example: *What is the same? What is different?* Ask learners to describe the shapes and introduce the names of the shapes, if appropriate.

Learners need to recognise and describe different patterns. Encourage learners to hold their pencils correctly and not to press too heavily on the paper when they are copying and making linear patterns.

I spy ... shapes

 Activity Book teaching notes

Pages 26–27

- Make sure that learners can distinguish the flat 2-D shapes on these activity pages from solid shapes. The pictures on page 26 are intended as an 'Explore' activity, so learners can talk about the shapes they see. Ask them what is special about a circle and look for circles in the classroom and in the picture. Ask them what is the same and what is different about the rectangle and the square.

- Give learners the opportunity to find objects that have the same shape, but other attributes that are different. They can look at the shapes on page 27 and join pairs that are the same shape. They may be inaccurate in their drawing of straight lines for shapes, but check they have the correct number of sides for each one.

Activity ideas

- Logic shapes: Provide the learners with a variety of flat 2-D shapes of different colours, sizes and thickness. Play 'dominoes' with the shapes where the touching shapes must have something in common. Learners must state the 'common feature' when placing their shape. For example: *It is the same shape. It is the same size. It is the same thickness.*

- Count-a-shape: Link shape activities to early counting skills. Ask: *Make me a line of five round shapes. Draw me a shape that has three spots.*

- Pass it on: Let learners sit in a large circle. Give them a flat shape to be 'passed on' round the circle. As the shape passes from learner to learner, a different fact should be given about the shape. Encourage a co-operative approach where everyone is helping to get the shape as far round the circle as possible.

Patterns

 Activity Book teaching notes

Page 28

- Give learners different patterned fabric or wallpaper to compare and talk about. Then let them compare and explore the patterns on page 28. Encourage them to use language that describes the different patterns accurately. Look for similarities and differences in the patterns.

- Make sure that learners can see the way a pattern repeats when they follow line patterns or create their own. Ask them what a single part of the pattern looks like.

Activity ideas

- Linear patterns: Set up trays of beads and laces for making patterns. Ask learners to work in groups, each with its own tray. Tell each learner to make a pattern. Look at the results and discuss: *What comes next? Tell me about the pattern. How many beads did you use? How will the pattern continue?* Swap the trays around and repeat.
- Printing: Ask learners to print with items such as cotton reels, sponge off-cuts, feathers, toothbrushes, and so on, to make a repeating pattern. Add these patterns to the class display or a book of patterns.
- Copy that: Let learners work in pairs to make a pattern by using items such as cubes, beads or pegboards. Then partners can swap and copy each other's patterns.
- Cyclical patterns: Encourage learners to make 'ring patterns' with apparatus such as beads or pegs on a pegboard. Discuss how the patterns go round and round; for example, ask: *Which bead is between the yellow ones?*

Solid shapes

 ## Activity Book teaching notes

Page 29

- Talk about the shapes that are used to make the model dragon on page 29. Ask learners to explain what makes each shape special and what makes some shapes the same or different from others. Learners could make a model dragon themselves from different junk materials.
- Show a model of a sphere, cylinder, cuboid and cone, and point them out on the page. Explain that there are shapes like these all around us. Show some real objects in these shapes for the children to explore and sort. The learners can then draw lines from the objects to the shapes on page 29.

Activity ideas

- Describing shapes: Talk about sorting shapes by similarities, using descriptions such as: *round like balls*; *with flat sides like this box*; *with curved and flat faces like this sponge*. Put something flat in one tray, and something round in another. Let learners choose an item, describe its shape and place it in the appropriate sorting tray. Play 'I spy', choosing two items that match for shape or texture.
- Venn diagram: Choose a criterion for sorting building bricks and solid shapes onto a single ring Venn diagram, without telling the learners what it is, for example, 'has corners'. Let learners take turns to place a shape on the Venn diagram, either inside the loop or outside. If the placing fits the criterion, say *yes*. Say *no* if it does not fit the criterion and give the learner time to reposition it appropriately. When all the shapes have been sorted, make a label for the diagram. Change the criterion for the sorting.
- Stacking and packing: Tell learners to 'stack' different shapes. Query which shapes are good for stacking. Ask learners to pack one type of shape into a container, such as balls into a box or bricks into a tin. Query which shapes pack well.

Success criteria

While completing the activities, assess and record learners who can:
- describe what flat shapes look like
- copy and draw different patterns
- use boxes to make models.

Assessment

- Ask learners to complete the *What can you remember?* activities on pages 30–31 of Activity Book A.
- Ask learners to self-check their understanding of key learning objectives covered this term using the self-assessment chart on page 32 of Activity Book A.

Learning objectives overview for Term 2

Units	Themes	Learning objectives	Activity Book pages	Preparation for Cambridge Primary Maths Stage 1 (Framework Code)
Unit 7 Numbers to 10	**Numbers to 10**	Read and represent numbers to 10.	4–5	1Nn1, 1Nn2
	Writing numbers	Write numbers to 10.	6	
	Matching numbers	Count and match objects to numbers to 10.	7	
Unit 8 Counting to 10	**Counting**	Count along a number track to 10.	8–9	1Nn1, 1Nn3
	Missing numbers	Count and know the position of numbers on a number track to 10. Use 'before', 'after', 'next' and 'middle' to describe the position of numbers on a number track.	10	
	Counting on	Start a count from any number to 10 on a number line.	11	
Unit 9 Addition and subtraction	**Finding totals**	Combine and count all the objects in two sets to make a total up to 10.	12–13	1Nc2, 1Nc8, 1Nc9
	Partitioning	Partition numbers to 10 in different ways.	14–15	
	Taking away	Subtract objects from a set of up to 10 objects and work out the number left.	16–17	
Unit 10 Time and position	**Story sequences**	Sequence everyday activities and events.	18	1MT1, 1Mt3, 1Gp1
	Months	Know some of the important months of the year, including birthdays and festivals.	19	
	Position of objects	Describe where objects are using positional words, for example, 'under', 'next to', 'over'.	20–21	
Unit 11 Measures	**Opposites**	Know and use the opposites of different comparative words. Compare the length or height of different items.	22–23	1Ml1, 1Ml2, 1Ml3
	Length	Compare and order the length or height of three or more items.	24	
	Ordering objects	Compare and order the size of three or more items. Compare and order the capacity of three or more containers.	25	

Units	Themes	Learning objectives	Activity Book pages	Preparation for Cambridge Primary Maths Stage 1 (Framework Code)
Unit 12 Shapes and patterns	Pictures and patterns	Make pictures and patterns with lines and shapes.	26	1Gs1, 1Gs2
	Flat shapes	Describe and name flat shapes. Compare properties of flat shapes, such as the number of sides.	27	
	Shapes around us	Compare shapes in the environment and recognise similarities and differences.	28–29	

Unit 7 Numbers to 10

 Activity Book B,
pages 4–7

 Story Book B

 Ten frames, page 75
Number cards 0–10, page 66

Learning objectives

- Read and represent numbers to 10.
- Write numbers to 10.
- Count and match objects to numbers to 10.

 Resources

beads, counters, straws, interlocking cubes, ribbons, ten frames, number cards to 10, small pieces of paper, red and blue pencil crayons, large pieces of card

Key words

count, match, numbers to 10, count on, count to, the same, number track, ten frame

Background information

Learners need opportunities to recognise the numbers 1–10 in different contexts, and to write the numbers and relate them to a matching number of items. Let them see the number 10 as 1 more than 9. At this stage, they do not need to understand the place value of a two-digit number.

Fingers are useful for counting and relating the last number in the count to how many. Make sure each item is counted carefully, with no items missed out or repeated. Learners may need to touch items as they count to help with their one-to-one correspondence: each number word matching each counted object in the set.

Numbers to 10

 **Activity Book teaching notes**

Pages 4–5

- Give the learners each a ten frame and some counters. Use PCM 10: Ten frames on page 75 of this Teacher's Pack. Ask them to match the frames on page 4 and then record the number they have made. Concentrate on the numbers 6–10, with the learners placing 5 counters on the ten frame to show 5, and then show 6 as 1 more than 5 and 7 as 2 more than 5, and so on. This will help them with counting on, rather than counting from 1 each time. 'Sets of 5' can then also be used as a strategy for addition in the future, when appropriate.

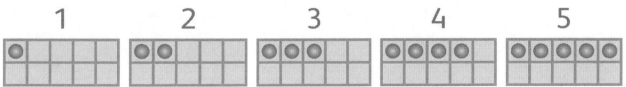

- Check the accuracy of learners' counting of the flowers in the second activity, with the movement of their finger from one flower to the next, matching the numbers when they say the number. As they write the numbers, encourage them to say each number aloud or silently in their head.

Activity ideas

- Count to check: Give learners counting objects, such as cubes, straws or counters, and a ten frame. Ask them to take a handful of the counting objects and guess how many they have, then count to check. They should arrange the items on a ten frame so they can check the count.
- Number hunt: Go on a number hunt and find different numbers around the environment, in books, on doors, on clocks and in the street. Numbers such as 4 and 9 can cause difficulties due to the varying styles in which they are used.
- Special number: Choose a special number for the day. Learners look out for that number as they go about their daily routines.
- Hide-and-seek numbers: Write the numbers 1–10 on small pieces of paper. Ask the learners to put them in different places around the classroom. Say a number and let learners go to find that number.

Writing numbers

 Activity Book teaching notes

Page 6

- Counting numbers of objects aloud goes alongside learning to be able to recognise numbers and then write the numbers. First model how the final number in the count is written.
- Use this page as practice in writing numbers, alongside recognition of the value of each number. Ask the learners to make 10 towers from interlocking cubes to represent the numbers 1–10. These can be put in sequence in front of them as they write the numbers.
- As they practise writing the numbers, encourage the learners to say the numbers aloud or silently in their head. Model the starting position for each number, with some numbers needing two strokes (4, 5 and 9) to complete the numbers.

Activity ideas

- Numbers in the air: Use two colours to write each number on large pieces of card. Learners stand and point to each number, tracing in the air first the red part, then the blue part. This will help them to see the different movements needed when writing numbers. Choose which numbers to introduce in this way.
- Behind the wall: Play the 'Behind the wall' activity by sliding a number card from 0–10 up from behind a 'wall'. A 'wall' can be a piece of card, a book or any other suitable screen. Have part of a number just peeping over the top of the wall. Ask: *Show me which number this might be. Could it be any other number? Show me.* Keep showing a little more of the number and repeating the instruction.

Matching numbers

 Activity Book teaching notes

Page 7

- Learners should recognise that the last number in the count is a label that shows how many items there are. Encourage them to count by touching each bee as they count. Learners need to count carefully, making sure they do not count the same item twice or miss any items out. Relate the last number in the count to the total.

- As they count the bees, the learners can place counters on a ten frame to match the count. This can then be used to reinforce the number with the last number of the count.

Activity ideas

- Show me: Each learner has a set of number cards 1–10. Use PCM 1: Number cards 0–10 on page 66 of this Teacher's Pack. Play 'Show me' activities where each learner shows a number card, by holding it up in the air. Hold up a large number card and ask: *Show me a number like this. What does it say?* Ask learners to hold up selected numbers: *Show me number 8. Show me number 6.*
- Number match: Play 'Show me' activities as above, but this time the learners match their number to a cube tower that you hold up. Count out the cubes together and then they show the number card that matches the total.
- Number match: Learners work in pairs, each learner having ten cubes and a set of number cards from 1–10. They take turns to show their partner some or all of their ten cubes. How quickly can their partner show how many there are with a matching number card?

Success criteria

While completing the activities, assess and record learners who can:

- show numbers to 10 on a ten frame
- write the numbers to 10
- count and match objects to a number card to 10.

Unit 8 Counting to 10

 Activity Book B, pages 8–11

 Story Book B

 Spinners, page 67
Number cards 1–10, page 66
Number track 1–10, page 70

Learning objectives

- Count along a number track to 10.
- Count and know the position of numbers on a number track to 10.
- Use 'before', 'after', 'next' and 'middle' to describe the position of numbers on a number track.
- Start a count from any number to 10 on a number line.

 Resources

0–3 spinner, beads, counters, cubes, washing line, pegs, number cards 0–10, ten frames, number track, number line, counting stick

Key words

count, match, numbers to 10, before, between, middle, after, count on, forwards, backwards, number track, ten frames, number line

Background information

The positional aspect of a number is literally its position in relation to other numbers, for example on a number track or number line. This positional aspect is important for the following reasons:

- It works well alongside the cardinal aspect of number. On a number track, five counters can be placed up to the number 5 and one more counter than this can be placed on the number 6.
- It supports the order of number words. Patterns can be seen and words practised in sequence – this is particularly the case for those tricky 'teen' numbers.
- It helps to develops language. Maths words such as 'before', 'after', 'next' and 'middle' can be explored.
- Most importantly, it helps to develop counting techniques: counting forwards and backwards, counting on and back from different starting points and counting in steps of different sizes.

Counting

 Activity Book teaching notes

Pages 8–9

- This counting game can be modelled and explored as a story before playing. Describe the adventures of a brave mountain climber who wants to get to the top of Mount Ten. Once at the top, he or she skies all the way back down again to his or her home at 'One'. Show that he or she goes back two squares because of the wind and one square because of the rain on page 8.
- Explain the rules of the game and demonstrate how to play: the slow climb up and the quick ski down, using a spinner to determine where to move to. Make a spinner using the template on page 67 of this Teacher's Pack, PCM 2: Spinners.

Activity ideas

- Dropping cubes: Drop ten interlocking cubes one by one into a tray (or other suitable resource) so each one makes a sound as the learners say the numbers in sequence. Ask five learners to come to the front. Give each learner a number card from 1–10 so they are standing in order. As the cubes are dropped, they should hold their card up to match the number. Use PCM 1: Number cards on page 66 of this Teacher's Pack.
- Boston wave: Learners sit in a circle and count slowly together: *One-two-three-four-five-six-seven-eight-nine-ten, one-two-three-four-five-six*. Decide which way round the circle the 'Boston wave' goes. As 'one' is chanted, point to a learner who quickly stands up, then immediately sits down. Point to the next learner in the circle as 'two' is chanted; he or she then also quickly stands up and sits down. Repeat the 'stand–sit' action going around the circle. With practice, learners will automatically continue the 'Boston wave' round the circle.

Missing numbers

 Activity Book teaching notes

Page 10

Use the washing line number track at the top of the page to count on and back as a class. Draw a track on the board (or use PCM 5: Number track 1–10 from page 70 of this Teacher's Pack) and play 'What's the number?' as a class. Use a picture cut out of a piece of card (such as a face), large enough to cover completely a number on your number track. Ask the learners to turn around or cover their eyes. Hide one of the numbers and ask the learners to say which number you are hiding. Ask them how they know, and talk about the numbers before and after the hidden number.

Activity ideas

- Washing lines: Peg the number cards 1 to 10 to a washing line in sequence. Each day, turn a number around and see if learners know the missing number. As a challenge, swap two of the numbers over and ask an individual learner to change them back to the correct position.
- Counting stick: Use a counting stick marked off in ten equal divisions. An unnumbered metre stick marked off in decimetres is ideal. One can be made by fastening coloured tape around a length of a broom handle. Label one end of the stick zero and the other 10. Move a finger slowly along the divisions, one at a time, counting in ones. Learners count in unison. Once learners become confident in counting forwards, include counting backwards along the stick.
- Hiccup counting: Include 'hiccup counting' on the counting stick. Stop part-way through the count sequence and move a finger back a division before continuing:
  one, two, three, four (move finger back); *three, four, five, six...*

Counting on

 Activity Book teaching notes

Page 11

- This is the first time learners have used a number line rather than a number track to represent the position and order of numbers. Model the use of a number line, jumping along from 0 to 10 in single steps.
- Demonstrate to the learners that you can count on from different starting numbers. Use a number line on a whiteboard and circle the number 4. Count on 1 and show the jump to 5. Repeat this, starting at 4 and counting on 2, 3, 4 and 5 while emphasising the jumps. Use the example at the top of page 11 to show the rabbit hopping from 3 to 7, counting on 4 from 3.

Activity ideas

- Washing lines: Peg the number cards 1 to 10 to a washing line in sequence. Ask learners to choose any starting number card to hold on to. Then ask another learner to count on 2 and to hold up that card. Repeat this so each learner takes a turn at counting on 2. Extend to counting on 3, 4 and 5.
- Counting stick: Use a counting stick marked off in ten equal divisions. Move a finger slowly along the divisions, one at a time, counting in ones. Learners count together. Put your finger on different starting points and ask the learners to count on 3. Once you have done this a few times, ask the learners to work out the end number each time before counting together.
- People numbers: Give out number cards in the range 0–10 and let the learners stand in line. Ask individual learners not holding a number card to change places with a given number. For example: *Gideon, change places with any number which is more than 7.*
  Seena, change places with a number between 2 and 6. Ask a learner to change places with someone who is 2 more than 6 and encourage them to start at 6 and count on 2. Repeat this for other numbers.

Success criteria

While completing the activities, assess and record learners who can:
- count forwards and backwards on a number track
- work out any missing numbers to 10 on a number track
- describe where a number is on a number track
- use a number line to count on from different numbers to 10.

Unit 9 Addition and subtraction

 Activity Book B, pages 12–17

 Story Book B

 Dominoes, pages 68 –69
Partitioning circles, page 76

Learning objectives

- Combine and count all the objects in two sets to make a total up to 10.
- Partition numbers to 10 in different ways.
- Subtract objects from a set of up to 10 objects and work out the number left.

 **Resources**

beads, counters, cubes, number track 0–10, number rods, dominoes, yoghurt pots, string, funnels, tubes, coins, box, squared paper, poster paper, small objects to partition

Key words

add, take away, combine, count, total, altogether, more, less, subtract, how many …?, how many more to make …?, how many more is … than …?, how many fewer is … than …?

Background information

Addition and subtraction are inextricably linked, the one being the inverse operation of the other. Activities should involve manipulating real sets of objects, with a focus on partitioning sets in a variety of ways to demonstrate number combinations.

Subtraction, as 'taking away', is easier for learners to understand if it is handled practically. For each example, use counters or cubes, if possible, to take an amount away from a starting number.

Finding totals

 Activity Book teaching notes

Pages 12–13
- These pages reinforce the specific addition structure of 'aggregation' – combining two or more quantities. This is linked to the language of 'How much/many altogether?' and 'What is the total?' Learners need to hear and use this language as they practically combine objects together to make different totals.

- On page 12, there is a number track to colour in two colours to match the number of objects in each set. Model the use of a number track with the learners, counting on from the first set to find the total.

Activity ideas

- Dominoes: Give small groups of learners a set of dominoes or use PCMs 3 and 4: Dominoes on pages 68 and 69 of this Teacher's Pack. Play 'sevens': all the touching dominoes must total seven. Change the 'touching' total.
- Combining machine: Learners make a combining machine in pairs, with two funnels or tubes leading into a box. They show two sets of cubes, such as 5 and 4, going into the machine. Ask them to say how many will come out of the machine and check by counting. Repeat for other amounts. Extend to using coins. Decide whether to include pennies only or other denominations as well.
- Under the pot: Each learner places five cubes under a small yoghurt pot (or plastic tub). They decide how many cubes they wish to take from under the pot to place on top. They may decide to take all of them. In turn, each learner says what they have, for example: *I have one and* (lifting the pot) *four*. They then change the number of cubes under and on the pot.

Partitioning

 Activity Book teaching notes

Pages 14–15

- Partitioning numbers in different ways is important for mental fluency and for learners to get a 'feel' for numbers, particularly with numbers to 5 and then to 10.
- Use cubes, counters or number rods to represent a start number and then ask learners to show different ways of partitioning or breaking up that number. Interlocking cubes are particularly useful as a model for this. Use PCM 11: Partitioning circles on page 76 of this Teacher's Pack for further practice of this concept.
- The + and = signs are not introduced at this stage. Instead '3 and 4 makes 7', or '8 is 3 and 5' prepares learners for the symbolic representation, and matches the physical actions involved in adding.

Activity ideas

- Adding with number rods: Ask learners to find different ways of matching a 6 rod with two other rods, such as a 4 rod and a 2 rod. They record by colouring on squared paper and writing a sum. Encourage putting the pairs of rods in an order and discuss whether all the possible solutions have been found. Repeat for a different starting rod.
- The story of 10: Put a large number 10 in the centre of a poster and ask learners to find different ways to make the total 10. Objects can be stuck on the poster or drawn to represent the different combinations.
- Partitioning a set: Each learner has a set of small objects on their workspace and a length of string with which to partition the set. They begin by counting the total number in their set (the number of objects in the sets may be the same or different for all learners). Each learner divides their set into two using the string. They then take turns to say how they divided or partitioned their set. For example: *I have put my seven into a two and a five. Eight can be shared into five and three.* After each partition they combine the sets together again and repeat with different partitions. Some learners may choose to include an empty set in their partitioning.

Taking away

 Activity Book teaching notes

Pages 16–17

- These pages focus on the early subtraction structure that involves 'taking away' objects. The principle is that there is a starting total and some are removed. However, there are a variety of strategies that learners may use to work out the number left over, including counting how many are left and counting back to find how many are left.

- It is always a good idea to reinforce the relationship between addition and subtraction, even though it is not the main focus for these pages.

Activity ideas

- Rod swaps: Each learner starts with a 7 rod (using number rods or similar). Show how 2 can be 'taken away' from the 7 rod by using 'fair swaps'. For example, changing the 7 rod for a 5 rod and a 2 rod. Show this as '7 take away 2 is 5'. Repeat with different rods, taking away 2 each time.

- Hiding items: Learners need seven small items and a plastic yoghurt cup. They take turns to hide some of the items under the cup. Ask: *How many items are hiding?* Change the number of items used.

- Bead thread: Learners use 8 beads and thread them on to a length of string. They then take away 3 beads by sliding them to one side and then count how many are left. Repeat with different numbers of beads.

Success criteria

While completing the activities, assess and record learners who can:

- put two groups of objects together and count the total
- break up 10 cubes in different ways and show the totals
- take away cubes from a set and say what is left.

Unit 10 Time and position

 Activity Book B, pages 18–21

 Story Book B

Learning objectives

- Sequence everyday activities and events.
- Know some of the important months of the year, including birthdays and festivals.
- Describe where objects are using positional words, for example: 'under', 'next to', 'over'.

Resources

story sequencing pictures, a camera, story books with story patterns, toys and small objects for positioning, simple calendars, plant pots, glove puppets, gym equipment such as tunnels and hoops

Key words

days of the week, months of the year, sequence, story, position, above, below, next to, near, between, under, over, through

Background information

Sequencing and putting events in order is an important step before telling the time and understanding the concept of the passing of time. Storytelling is a good opportunity for this, so make the most of the stories you read together, asking the learners to recall the order of events and describing what comes before or after a scene in the story.

Learning the names and the order of months in the year takes time, but a learner's birthday is a good starting point. Ask them to learn the months before and after their birthday. This can slowly build up their recall of the order of the months.

Positional language can be confusing for learners if they are not involved practically in tidying, sorting and arranging toys, clothes, and so on. Use the language accurately and encourage the learners to use key positional words during play activities.

Story sequences

 Activity Book teaching notes

Page 18

Before asking the learners to put the pictures in sequence, talk about each of the stories. Ask the learners what is happening in each picture. Then ask what they think might happen first. They can describe what happens next and then describe the events in the final frame of each scene. Ask them to point out the 'clues' that show the sequence of events.

Activity ideas

- Story patterns: Read stories with a strong sequence of events, such as:
  - The Very Hungry Caterpillar by Eric Carle (Puffin)
  - *The Zoo in our House* by Heather Eyles and Andy Cooke (Walker)
  - *The Enormous Turnip* by Katie Daynes (Usborne)
  - *A Hard Day's Work* by Mick Gowar (Scholastic)

  Ask learners to predict or recall the next event in the story. They could make pictorial representations of some aspect of the story.

- Photo puzzle: Take photos of a learner in the process of building a model from Lego, or completing a jigsaw puzzle. Take four photos from start to finish, print and cut them out as sequencing cards. Ask learners to put the cards into the correct order.

Months

 Activity Book teaching notes

Page 19

- The learners may not be able to read these months, but they can join in saying the months together and perhaps recognise the initial letters of each. It is through remembering important months in their own lives that they will build up the sequence of the months, so focus on 'special' months and then let them learn the months before and after these.
- Talk about the special events in learners' lives before they draw an event in the circle on page 19. They then join this event to the matching month.

> ### Activity ideas
>
> - Birthday song: Teach learners the following birthday song to the tune of 'Frère Jacques':
>   *One year older, One year older,*
>   *5 today, 5 today,*
>   *Have a happy birthday,*
>   *Have a happy birthday!*
>   *Celebrate! Celebrate!*
>
>   During circle time, sing the birthday song for those learners who have a birthday that day or week. Encourage the learners to clap once for each year of the learner's life.
> - Birthday chart: Make a class birthday chart. It might:
>   - show who has a birthday in each month
>   - record birthdays as they occur throughout the year
>   - be a timeline for the school year, with the learner's birthdays marked, in order of occurrence.
> - Get in order: Discuss the order of birthdays. Decide who is the oldest and the youngest in the class and then put the rest of the class in order for each month between them.

Position of objects

 Activity Book teaching notes

Pages 20–21

Talk about the picture across pages 20–21. Encourage learners to describe the different toys and to explain where they are on the shelves. Questions are provided as starting points and others can be asked using similar positional language.

> ### Activity ideas
>
> - Gym time: Use large pieces of apparatus to develop the language of position. For example, ask the learners to climb up high, crouch down, go in/out of tunnels, go through/over/under different gym equipment and turn left/right/move far away or come near. Ask learners to find different ways to go under and go through. Use hoops to climb through, jump in, and stand inside in pairs.
> - Plant pots: Line up some small toys in front of a large upside-down plant pot and get learners to count them. Move a few toys to the side of the plant pot. Ask: *How many in front? How many at the side? How many altogether?* Place some on top of the plant pot. Ask: *How many in front? How many on top? How many at the side? How many altogether?* Hide some toys underneath the plant pot. This introduces learners to the idea of 'missing numbers'. Ask: *How many in front of the plant pot? How many under the plant pot? How many altogether?* Discuss ways of finding out the number of toys hiding under the plant pot.
> - 'Simon says' game: This game can be played with learners controlling a glove puppet and is useful for encouraging understanding of position words. One learner stands at the front of the class, holding the puppet and you give them instruction. If the learner hears 'Simon says', he or she performs the action. If the instruction does not start with 'Simon says', the learner stays still. Learners who move incorrectly are out and they then pass the puppet to the next learner. For example:
>   *Simon says: put the puppet up in the air.*
>   *Simon says: put the puppet behind your back.*
>   *Simon says: put the puppet on your knee.*
>   *Pick it up again ... (learner does not move for this instruction!)*

Unit 11 Measures

 Activity Book B,
pages 22–25

 Story Book B

Learning objectives

- Know and use the opposites of different comparative words.
- Compare the length or height of different items.
- Compare and order the length or height of three or more items.
- Compare and order the size of three or more items.
- Compare and order the capacity of three or more containers.

 Resources

hoops, cubes, small objects for comparing size and length, strips of scrap paper and large square paper, stacking and nesting toys, building bricks, number rods, interlocking cube number rods, feely box

Key words

opposite, compare, same, different, length, height, order, capacity, full, empty, size, small, big, large, smaller, larger, bigger, long, longer, short, shorter, longest, shortest, tall, taller, tallest

Background information

The use of comparative language and opposite terms (such as 'tall/short') are important principles as preparation for ordering and then measuring length and size. They also prepare the way for the idea of inverses involving quantities. For example, when two quantities are compared, there are two ways of expressing them: *A is greater than B* and *B is less than A*. This follows the same comparative structure as, for example, *A is bigger than B* and *B is smaller than A* or *A is wider than B* and *B is narrower than A*. Encourage learners to make both equivalent statements when comparing objects. They will then be looking at differences between objects, which will help them identify equivalences when it comes to measuring. *What is the same?* and *What is different?* are key questions that focus on the concepts of transformation and equivalence – both fundamental ideas in mathematics.

Opposites

 Activity Book teaching notes

Pages 22–23

- These pages give learners the opportunity to explore pairs of opposites. This supports the idea of comparing objects and develops language associated with comparison and measures.

- Encourage learners to make both equivalents opposite statements when comparing objects. As they complete each activity of joining opposites or drawing opposites, ask the learners to say the opposite words, such as *short* and *tall*. If they are 'measures' words, then let learners use them in two equivalent sentences, for example: *This flower is taller than this one.* and *This flower is shorter than this one.*

Activity ideas

- Body opposites: Ask learners to explore opposite movements. For example:
  - quick and slow movements; stretch up high, curl up low
  - climb up, climb down; make a wide shape, a narrow shape
  - stand in a large hoop, in a small hoop.

  Let them use language of position such as *front*, *back*, *left*, *right*, *up* and *down*.

- Sizes: Use everyday classroom items, such as pencils, books and crayons. Ask learners to find pairs of opposites and to say how they are opposite, for example:
  - *This book is big and this one is small.*
  - *This pencil is long and this one is short.*

- 'Simon says' game: Play the 'Simon says' game (see page 40 of this Teacher's Pack) using opposites: *hands up high, hands down low; hands in pockets, hands out of pockets.*

- Construction sets: Use different types of construction sets and ask learners to build:
  - a tall and a short tower
  - a long and a short roadway
  - a big and a little monster.

Length

 Activity Book teaching notes

Page 24

Before comparing the length of the pictures of objects on page 24, give learners sets of different objects to compare lengths, such as pencils, belts and ribbons. Ask them to look for similarities and differences. They must then use the language of comparison to describe and compare them and identify a single item. Ask: *Which is largest? Longest? Smallest? Shortest?*

Activity ideas

- Comparing strips: Give each learner a strip of scrap paper. The strips can be different lengths. Ask: *Who has the longest/shortest strip? Which strips are about the same length? How can the strips be made the same length?* Explain how to find the half-way point along strips by folding.

- Comparing names: Use strips of large squared paper. Ask each learner to write their first and last name on the squared paper, one letter per square. Ask: *Does your first name have more or fewer letters than your second name? How many more? Who has a name where both names have the same number of letters?* Compare the lengths of pairs of names. Ask: *Who has the longest name? Who has the shortest name?*

Ordering objects

 Activity Book teaching notes 43

Page 25

There are three distinct parts to this page, with opportunities for the learners to order by length, by size and then by capacity. It is a good idea to do each part separately, with practical activities given first so they learn how to order three objects by comparing one object with another and then comparing a third (principle of transitivity).

Activity ideas

- Stacking and nesting: Use toys that stack or nest in order of size, for example stacking beakers and boxes. Discuss order and position. Use language of comparison such as *bigger, biggest*.

- Building bricks: Ask learners to make a tower of bricks with the largest brick at the bottom and the smallest at the top. They then make one tower taller or shorter than the other and then a third tower of a different height. Ask them to put their towers in order of height.

- Cinderella: Read the story of Cinderella to your class and use the opportunity for learners to compare their foot sizes. They draw around their feet and cut out the shapes. Ask learners to lay one foot outline on top of another, so that a direct comparison of size can be made. The foot outlines can be put in order of size and stuck onto paper as a record.

- Number rods: Use number rods or interlocking cube number rods:
  - Discuss ordering and comparing the rods.
  - Explain how two or more rods can match the length of other longer rods.
  - Make step patterns with the rods and discuss 'one more'.
  - Place one of each rod inside a feely box. Ask learners to find a rod that is longer/shorter/the same as the one you show them.

Success criteria

While completing the activities, assess and record learners who can:
- say the opposites of words such as *open, down, big, tall* and *empty*.
- find things longer or shorter than their pencil
- put objects into order of length from the longest to the shortest
- put objects into order of size from smallest to largest
- put four jugs in order of capacity, from full to empty.

Unit 12 Shapes and patterns

 Activity Book B,
pages 26–29

 Story Book B

Learning objectives

- Make pictures and patterns with lines and shapes.
- Describe and name flat shapes.
- Compare properties of flat shapes, such as the number of sides.
- Compare shapes in the environment and recognise similarities and differences.

Resources

2-D shape tiles, 3-D solid shapes, patterned fabrics and wallpaper, beads, laces, cubes, 3-D interlocking shapes, geometric shapes, paper, crayons/markers

Key words

flat shapes, solid shapes, circle, triangle, square, rectangle, cube, cuboid, cone, cylinder, face, side, points, corners

Background information

Talk about patterns around the school and home. Look at tiles, bricks, plants, fabric, and so on. Encourage learners to look at repeating patterns of colour and shapes, aiming for them to be able to recognise, extend and create patterns using drawings and concrete materials.

Your learners will be starting to recognise and name certain flat (2-D) shapes and solid (3-D) shapes. Talk about their properties, such as the shape of the faces of solid shapes and the number of sides of flat shapes. As they colour and match shapes, ask them to talk about any similarities and differences between the shapes.

Pictures and patterns

 Activity Book teaching notes

Page 26

- This page has a focus on some of the different attributes that repeating patterns can be generated from – colours, shapes and types of line. Talk about the three patterns in the 'Learn' section at the top of the page and ask what they notice about each of them. Learners can then work with their partners to decide what will come next in the pattern, and continue it for three or four more shapes, colours and line patterns.

- For the patterns on this page, the learners trace and draw shapes, which is a closed activity. They can then colour their patterns in any way they choose. Talk about the colours they decide on and ask how they make repeating patterns.

Activity ideas

- Linear patterns: Set up trays of beads and laces for making patterns with two criteria – shape and colour. Learners work in groups, each with its own tray. Tell each learner to make a pattern using colours and shapes. Look at the results and discuss: *What comes next? Tell me about the pattern. How many beads did you use? How will the pattern continue?* Swap the trays around and repeat.

- Pattern mats: Each learner has a sheet of paper and a selection of crayons or markers. Learners follow instructions from you for drawing different types of line down the sheets: *Draw a zigzag line. Draw a straight line. Draw a wavy line. Draw loops. Draw double lines. Draw dotted lines. Draw shaky lines.* Extend to thick and thin lines.

- Shape patterns: Make simple repeating patterns from interlocking cubes, 3-D interlocking shapes and geometric shapes. Ask learners to predict subsequent shapes in the sequence. Use colour, shape and size attributes.

Flat shapes

 Activity Book teaching notes

Page 27

- Give learners shape tiles to handle and look at, and compare them to the pictures of the four shapes in the 'Learn' section at the top of the page. Ask the learners if they know the names of the shapes. Repeat the names so they can learn them and pronounce them correctly. Talk about the similarities and differences between the shapes and focus on the number of sides of each. Explain that a circle has one curved side while the other shapes have straight sides.

- Explain that shapes of objects around us may not be exactly square, for example, but we would still say that it is 'shaped like a square'. Use that language when looking at the shapes on page 27 and show learners how to match them to the shapes in the centre of the page.

Activity ideas

- Find a pair: As a class or in a group play 'Find a pair'. Have a selection of different size and colour shape tiles – triangles, squares, rectangles and circles. Learners select a shape tile and find someone else with the same shape. They sit down in pairs, until all are paired up.

- What's my shape: Use a simple wall made of card. Slowly slide up a flat shape until a small part of it is showing above the wall. Learners select a shape from a pile of shape tiles they think it might be. Encourage them to talk about the shapes. Gradually show a bit more of the shape and let them decide to keep their first shape or change it for a different shape until it is completely shown.

- Pick up pairs: In small groups of 3–4, play 'Pick up pairs'. Lay out flat shape tiles and ask learners to take turns to pick up matching shapes. If pairs of shapes match, they keep both shapes; if pairs do not match, they are returned to the table.

Shapes around us

 Activity Book teaching notes

Pages 28–29

- Ask the learners to find shapes around them that are shaped like boxes and those that are shaped like balls. Talk about similarities and differences and what makes each type of shape special (rolling and stacking are two attributes, for example).

- Show models of a sphere, cylinder, cone, cuboid, pyramid and cube, and point them out on page 29. Explain that there are shapes like these all around us. Show that some shapes roll and some have flat faces, so do not roll easily. Show some real objects in these shapes for the learners to explore and sort by rolling/not rolling. The learners can then colour the shapes that roll on page 29.

Activity ideas

- Sorting toys: Use four sorting trays each with a different solid shape: cube, cuboid, sphere and cylinder. Ask learners to find everyday objects and toys and place them in the correct tray to match the shape. A toy may have only one part the correct shape, for example, the trailer on a tractor may be a cuboid.

- Building shapes: Learners use a variety of wooden or plastic solid shapes to manipulate and explore. Encourage them to join shapes together to make new shapes and build with the shapes. Ask them to build with only one type of shape: *What can you build if you only use cubes?* Talk about the shapes that are easy to stack and ask why they think they can stack well.

- Feely bag: Put four solid shapes on the table in a row: cube, cuboid, sphere and cylinder. Prepare a 'feely' bag with matching 3-D shapes inside. Learners take turns to put their hand into the bag without looking, and 'feel' a shape. When they are ready, they point to the same shape on the table, taking the shape out of the bag to see if it matches. Ask about the shape they are feeling: *Is it round? Is it curved? Does it have flat sides? Does it have points (or corners)?*

Success criteria

While completing the activities, assess and record learners who can:

- make and draw patterns with lines and shapes
- point to squares, triangles and circles that are around them
- talk about the number of sides of flat shapes
- say what is the same and what is different about shapes around them.

Assessment

- Ask learners to complete the *What can you remember?* activities on pages 30–31 of Activity Book B.
- Ask learners to self-check their understanding of key objectives covered this term, using the self-assessment chart on page 32 of Activity Book B.

Learning objectives overview for Term 3

Units	Themes	Learning objectives	Activity Book pages	Preparation for Cambridge Primary Maths Stage 1 (Framework Code)
Unit 13 Numbers to 20	Counting on from 10	Count on from 10 to 20.	4–5	1Nn1, 1Nn2, 1Nn8, 1Nn9
	Counting to 20	Read numerals to 20. Count and know the position of numbers on a number track to 20.	6	
	Numbers to 20	Write numbers to 20. Count a given number of objects from a larger set up to 20.	7	
Unit 14 Grouping and sharing	Counting groups	Put objects into equal groups of 2, 5 or 10 and count the groups and totals.	8–9	1Nc19, 1Nc20, 1Nc22
	Doubles	Put sets of objects of the same number together and relate to doubling.	10	
	Sharing	Share objects equally between two and relate to halving.	11	
Unit 15 Addition	Making totals	Partition numbers to 10 in different ways.	12–13	1Nc1, 1Nc8, 1Nc11
	Counting on	Count on from a number on a number track to 10 to add numbers together. Combine and count all the objects in two sets to make a total up to 10, counting on from the largest set.	14–15	
	Addition stories	Complete addition facts with missing numbers. Use addition facts to answer word problems.	16–17	
Unit 16 Subtraction	Finding the difference	Find the difference between two lines of cubes by comparing and counting.	18	1Nc1, 1Nc9, 1Nc11
	Counting back	Count back from a number on a number track to 10 to take away a number.	19	
	More counting back	Count back from a number on a number track to 10 to take away a number.	20	
	Hiding objects	Solve subtraction problems involving a 'hidden' number of objects.	21	

Units	Themes	Learning objectives	Activity Book pages	Preparation for Cambridge Primary Maths Stage 1 (Framework Code)
Unit 17 Measures and time	**Measuring length**	Use uniform non-standard units such as cubes to measure lengths.	22–23	1Ml1, 1Ml2, 1Ml3, 1Mt1, 1Mt2
	Weight and capacity	Use a balance to compare the mass of objects. Compare capacities of different containers and check by pouring.	24	
	Time	Recognise some hour times on analogue and digital clocks.	25	
Unit 18 Shapes and patterns	**Flat shapes**	Name flat shapes, describing some generalised properties of each shape. Identify shapes in different positions and orientations. Make repeating patterns with shapes.	26–27	1Dh1, 1Gs1, 1Gs2, 1Gs3
	Solid shapes	Recognise the face shapes of solid shapes. Name some solid shapes and describe properties.	28	
	Symmetry	Recognise simple shapes and objects that show reflection and symmetry.	29	

Unit 13 Numbers to 20

 Activity Book C, pages 4–7

 Story Book C

 Number track 1–20, page 72
Number lines 0–20, page 73

Learning objectives

- Count on from 10 to 20.
- Read numbers to 20.
- Count and know the position of numbers on a number track to 20.
- Write numbers to 20.
- Count a given number of objects from a larger set up to 20.

Resources

small toys or items, hoops or string, small box and bag, beads, counters, interlocking cubes, washing line, pegs, number cards 0–20, ten frames, number track, number line, counting stick

Key words

count, match, numbers to 20, before, between, middle, after, count on, forwards, backwards, number track, ten frame, number line

Background information

The 'teen' numbers often cause difficulties because they do not follow the written or spoken pattern of numbers beyond twenty. The place value of the numbers is less important at this stage than knowing the order of the numbers, reading them accurately and using them to count.

11 and 12 can be introduced as 'one more' and 'two more' than 10. Encourage the learners to count in groups of 5, so 11 is '5 and 5 and 1'. 'Teen' numbers are easier to visualise if learners can recognise 10, and breaking 10 down into '5 and 5' is a useful step towards this.

Counting on from 10

Activity Book teaching notes

Pages 4–5

- Learners need to use small toys or items to make numbers 11 to 20. Demonstrate putting 10 toys in two rows of 5, so they can quickly count 5 and 5 to make 10. Put a small hoop or piece of string around these to show a group of 10 and then put 1 toy outside the group. Show that this is 10 and 1, which is 11. Continue this with numbers to 20.

- Use these pages to talk about the lines of cubes showing 10 and, for example, 6 more for 16. The learners use interlocking cubes to show different numbers, counting on from 10 each time to say the number.

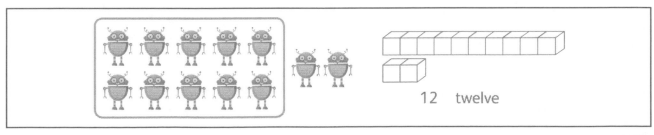

12   twelve

Activity ideas

- Number steps: Ask learners in groups to use interlocking cubes of two colours, with one colour used to make 10 columns of 10 cubes. These are arranged and lined up next to each other. Then the other colour cubes are added on to each column to show the numbers 11, 12, 13, 14, and so on, to 20. Talk about the patterns they can see with the completed set of steps, and then pick out an individual step for learners to count on from 10 to say the number.
- Tap the box: Put 10 counters in a small box and write the number 10 on top of the box. Put 10 more counters in a bag. Let individual learners take out a small handful of counters to place next to the box. Ask how many counters there are altogether, including those in the box. Encourage them to tap the top of the box as they say 10, and then count on as they touch each individual counter to reach the total.

Counting to 20

📄 **Activity Book teaching notes**

Page 6

- Count forwards and backwards together along the number track in the 'Learn' box at the top of the page, asking the learners to touch each number as they say it aloud. Repeat from different starting numbers, going forwards and backwards.
- Use a large number track for all the learners to see and include some 'hiccup counting': stop part-way through the count sequence and move a finger back a division before continuing: *eleven, twelve, thirteen* (move finger back) *twelve, thirteen, fourteen…* You could use an enlarged version of PCM 7: Number track 1–20 (on page 72 of this Teacher's Pack) or PCM 8: Number lines 0–20 (on page 73 of this Teacher's Pack) for this activity.

Activity ideas

- Number cards: Ask learners to line up holding number cards 10–20 in sequence, without using a number track if they can. They then check the order to see if it matches the track. Ask about the position of the numbers: *What number comes before 16? What number is after 13?*
- People numbers: Give out number cards in the range 0–20 and stand the learners in line. Ask learners not holding a number to change places with a given number. For example: *Zara, change places with any number that is more than 7. Ishaq, change places with a number between 12 and 16.* Turn some of the learners holding numbers around so that their backs face the rest of the class. This makes a 'missing numbers' sequence. Choose learners to go and tap one of the 'missing numbers' on the back and to say: *I think you are number …* If they are correct, they change places with that learner/number.
- Number cards: Ask some learners holding number cards 0–20 to line up in a muddled order. Ask learners without cards to give instructions so that the learners with number cards become ordered correctly.

Numbers to 20

📄 **Activity Book teaching notes**

Page 7

- This page gives an opportunity for learners to practise writing the numbers 11–20. Show the starting positions for writing each number and encourage learners to say each number as they write it.

● Make sure they understand that the '1' written first in each number to 19 represents 10. Being able to see this as 10 and count on from 10, rather than the concept of place value, is the focus. If this is still causing difficulties, use some of the activities from the previous pages with extra support in recognising, for example, 14 as '10 and 4 more'.

Activity ideas

● Number cards: Ask learners in small groups to place number cards to 20 (could be limited to numbers 10–20), upside-down on the table. In turn, each learner picks up a card without showing the others, and uses buttons, counters or cubes to match the same amount as the number on the card. The other learners count and say the number, then check it is correct. They then place the card face down beside the set of objects.

● Pick up pairs: Play 'Pick up pairs' in small groups. Two sets of 11–20 number cards are mixed up and spread around randomly, face up so the numbers can be seen. Learners take turns to select two cards that have the same number and say the number name at the same time. If they are a matching pair, they keep the cards. This continues until all the cards have gone. Once completed, the learners write the numbers they have collected.

Success criteria

While completing the activities, assess and record learners who can:
● count on from 10 to reach any number to 20
● read the numbers to 20
● describe where a number to 20 is on a number track
● write the numbers to 20
● count up to 20 objects and say the total.

Unit 14 Grouping and sharing

 Activity Book C, pages 8–11

 Story Book C

 Number track 1–20, page 72
Partitioning circles, page 76

Learning objectives

● Put objects into equal groups of 2, 5 or 10 and count the groups and totals.
● Put sets of objects of the same number together and relate to doubling.
● Share objects equally between two and relate to halving.

 **Resources**

small items (buttons, marbles, interlocking cubes, function machine, counters), beads, counters, cubes, number cards 0–10, paper plates, number track, number line, counting stick

Key words

groups, equal, total, altogether, double, share, halve, count, match, sets, numbers to 20

Background information

Learners often have difficulties with the concepts of multiplication and division because of the complex structure of each, along with the associated language. There is a network of connections within and between them, with interpretations that could be based on all or any of repeated addition (and subtraction), scaling, arrays, sharing or grouping.

Early multiplying involves equal grouping of objects and counting the groups (linked to repeated addition), so the language is '3 lots of 5' or '4 groups of 2'. This will lead on to the use of the multiplication or 'times' symbol ×, which can replace 'lots of' or 'groups of'.

Division in this unit has a focus on equal sharing with the key language of 'share equally between'. This is often the first experience of division for young children, but it is not the only structure and is not the model used for representing the ÷ symbol. An important idea for learners to understand in their next stage will be the inverse of multiplication based on grouping, for example how many groups of 3 make 12, which links to '12 ÷ 3 = 4'.

Counting groups

 Activity Book teaching notes

Pages 8–9

- It is important that learners know that they are making *equal* groups of objects and then counting the groups for these activities. They then work out how many there are altogether by counting on.
- Repeat the language for each activity so it follows the same pattern: *How many in each group? How many groups? How many altogether?*

Activity ideas

- Equal grouping: Ask learners to count out 12 cubes onto their workspace. Ask them to put the cubes into twos. Ask: *How many groups of two are there? What if we made groups of 3? Put out 15 cubes. How many groups of 5 are there?* Repeat for other amounts and groupings.

- Number track match: Learners use a number track and cubes. They make sticks of two cubes and find out how many groups of two are needed to reach 12. They record this as a sentence: *There are 6 twos in 12.* Use PCM 8: Number track 1–20 (on page 72 of this Teacher's Pack) for this activity. This can be repeated for different amounts and then for different groupings.

- Spotting groups: Set up quantities of small toys in separate groups of 2, 5 and 10. Ask: *I want to be able to give a group of 5 learners a toy each. Where is the group of five?* Look for learners who can spot the group without counting the toys individually. Repeat with the other groups.

- Groups of items: Use a set of 20 small items (buttons, marbles, cubes or counters). Ask learners to put them into groups of 10, then 5, then 2. Ask them how many groups there are for each, and how many in each group. Ask if they can work out how many items there are altogether.

Doubles

 Activity Book teaching notes

Page 10

- Talk about the dots on the dominoes in the 'Learn' box at the top of the page, and ask if learners can see anything that is special about them. Ask if they know what 'double' means and share responses. Draw dominoes on the board with dots on one side and blank on the other. Ask learners to draw in dots to make doubles and repeat the numbers, for example, 'double 3 is 6'.

- The learners use interlocking cubes to make pairs of doubles. They put them together to make rectangle arrays such as 5 and 5 to make 10. The L-shapes show odd numbers that fit together to make an even number, which some learners may notice.

Activity ideas

- Doubling machine: Make a function machine from a large box, so there is an IN and OUT and a label on the box with DOUBLE written on it. Put a tray of cubes inside the machine and ask learners to use the machine to double their towers. Give a learner a tower made from 3 cubes. They then put it through the machine, make another tower of 3 cubes to match and then combine them to make a tower of 6 cubes. Compare the two towers, showing that 6 is double 3, breaking the 6 tower in half and putting it together to demonstrate this. Repeat this for different-sized towers.

- Towers: Prepare some cube towers with two sets of each length, up to 5 cubes. Ask the learners to choose pairs that are the same length and ask how many cubes there are altogether. Give them the 3-cube tower and asking them to make this double the length. Repeat with other lengths, asking how many cubes there are in each length and what double that number is.

Sharing

 Activity Book teaching notes

Page 11

- This page introduces sharing between two to represent halving as the inverse of doubling. Sharing equally is the important idea here, so check that the learners understand this first. Use the diagram on the activity page that partitions counters into two equal groups as the practical model for halving.

- If appropriate, show the inverse with counters moving from the circle into the square to show doubling.

Activity ideas

- Paper plate shares: Learners use paper plates and cubes. They count out some cubes and share them between two plates, recording the sharing as a sentence. They choose different amounts and repeat the activity.

- Halves and doubles: Learners place the same number of cubes on each side of a circle that has been partitioned in half. Talk about the facts they can say about this model. For example: *4 and 4 makes 8; Double 4 is 8; Half of 8 is 4; 2 groups of 4 makes 8.* Use PCM 11: Partitioning circles on page 76 of this Teacher's Pack for this activity.

- Tower snap: Learners make a tower with cubes and count how many cubes they used. They then snap their towers in half, compare each half, adjust for fair shares and record their sharing: *6 makes two sets of 3; 7 makes two sets of 3 and has 1 left over.*

Success criteria

While completing the activities, assess and record learners who can:

- put objects into equal groups and count the groups and totals
- double and halve small sets of objects
- share objects between two people.

Unit 15 Addition

 Activity Book C,
pages 12–17

 Story Book C

 Number cards 0–10, page 66
Spinners, page 67
Number lines 0–10, page 71

Learning objectives

- Partition numbers to 10 in different ways.
- Count on from a number on a number track to 10 to add numbers together.
- Combine and count all the objects in two sets to make a total up to 10, counting on from the largest set.
- Complete addition facts with missing numbers.
- Use addition facts to answer word problems.

Resources

beads, counters, interlocking cubes, number track 0–10, number rods, dominoes, yoghurt pots, string, number balance equaliser and weights, pegs, number cards 0–10, squared paper, paper strips

Key words

more, more than, most, greater, addition, add, makes, altogether, partition, combine, total, count on, equal to

Background information

There are two main structures for addition explored in this unit: counting on to increase a number and the union of two sets (combining). Counting on relates well to the ordinal aspect of number, using number lines and number tracks to model the addition. Combining sets of objects together to make a total links well with the language of addition and to the + symbol when that is introduced. 'How many altogether?' is the key question that is asked when combining sets of objects.

Making totals

 ### Activity Book teaching notes

Pages 12–13

- The + and = symbols are introduced on these pages, so spend some time talking about them and modelling their use practically. The additions on each page can be modelled using counters or cubes.

- Make sure learners understand the = symbol as 'is equal to' rather than 'makes the answer'. Writing 5 = 3 + 2 and 3 + 2 = 5 will help them see this as a balance, with one side of the = symbol the same total as the other side.

Activity ideas

- Cube trains: Give a train of 3 cubes to learners. Give them 6 more cubes as a train and ask them to work out the total. Ask them to explain as they are working it out and observe the strategies. Do they count on from the largest number? Repeat for other numbers.

- Partitions: Give learners a number between 5 and 10. Using two different colours of interlocking cubes, they make as many different towers as they can, exploring how to partition a number in several ways. They talk about the different ways they can make the number and record their results by colouring squares on squared paper. As a challenge, they can include cubes of three colours and show the results.

- Make 5: Use number cards from 0 to 5 from PCM 1: Number cards 1–10 on page 66 of this Teacher's Pack. Shuffle them up and turn them over one at a time for learners to respond quickly with the number pair that totals 5.

- Equaliser sums: Provide a number balance equaliser and three weights. Learners hang the three weights on the equaliser to find different ways of making it balance. They record this as a calculation using symbols.

Counting on

 Activity Book teaching notes

Pages 14–15

- The number track in the 'Learn' box at the top of page 14 shows how to count on 3 from 5. Demonstrate this and show how it is also represented by adding the marbles together. Count on the number track with other starting positions and additions so the learners understand the method.

- The game can be played by up to 4 players. Use PCM 2: Spinners on page 67 of this Teacher's Pack. Demonstrate how to spin this once to find the starting number and spin again to count on that number of places.

Activity ideas

- Show me: Ask learners to put a counter on the number 4 on a number line 0–10. Use PCM 6: Number lines 0–10 on page 71 of this Teacher's Pack. Say: *Count on to show me 3 more.* Learners move another counter from the start position to show how they count on, and explain how they are doing this. Repeat for other start numbers and jumps.

- Maths washing line: Fasten numbers 0–10 to a washing line with pegs. Put rings round two of the pegs, such as 2 and 7. Discuss counting on from 2 to 7. Repeat for other numbers. Change it so you put a ring on a single number, for example around 3. Ask the learners to choose any number to 5. An individual learner then counts on that number from 3 and puts on another ring.

- Number track add: Learners work in pairs. They shuffle number cards 1 to 5 and turn over two cards. Each learner makes a tower from cubes to match the number on the card. They combine their towers in a number track and write an addition sentence. Repeat for other pairs of cards.

Addition stories

 Activity Book teaching notes

Pages 16–17

- These pages give practice in adding in a context and 'missing number' problems. With missing number additions, make sure the learners understand the format, and the importance of balancing either side of the = symbol.

- Encourage learners to talk about the additions as stories, acting them out or telling the stories to each other. Use practical resources to help learners make sense of them.

Activity ideas

- Missing number additions: Use up to 10 small items and a yoghurt pot. Ask learners to count how many items there are and then cover 3 items with the yoghurt pot. Ask how many they can see and how many are hidden. Repeat for different numbers. Show these as missing number additions, for example, 7 + __ = 10.

- Maths stories: Give learners objects that they can use to act out and solve word problems. These objects can be coins, counters, cubes or small toys. Give additions as stories for learners to act out and solve. For example: *Four frogs sit on a leaf. Three more join them. How many frogs altogether? You have 2 dollars. An extra 6 dollars is given to you. How much do you have now?*

- The story of…: Provide some number rods. Ask learners to find different ways of making the story of a number, such as 8. They can record this by using symbols. Check whether a number pattern is used or random pairs found by trial and error.

- Cover up: Learners work in small groups with two sets of number cards 1 to 5 laid face down on a table. The learners each write the numbers 1–10 on strips of paper. Turns are taken to turn over two cards and total them. Numbers matching the score are crossed off that player's number strip. For example, if 7 is scored, then 7, 6 and 1, 5 and 2, or 4 and 3 can be crossed off.

Success criteria

While completing the activities, assess and record learners who can:
- make numbers to 10 in different ways and show them as an addition
- use a number track to add by counting on
- put two groups of objects together and count on from the largest group to find the total
- work out the missing numbers in an addition
- make up addition stories.

Unit 16 Subtraction

 Activity Book C, pages 18–21

 Story Book C

 Numbers lines 0–10, page 71
Spinners, page 67
Number track 1–10, page 70

Learning objectives

- Find the difference between two lines of cubes by comparing and counting.
- Count back from a number on a number track to 10 to take away a number.
- Solve subtraction problems involving a 'hidden' number of objects.

Resources

beads, counters, interlocking cubes in different colours, number track 1–10, number rods, dominoes, yoghurt pots, 1–3 spinners, number lines 0–10, string, pegs

Key words

difference, take away, less, subtract, how many fewer is … than …?, count on, count back, number track, number line

Background information

Early subtraction involves partitioning and taking away objects. The basic idea is that there is a set of objects and an amount is taken away. There are a variety of strategies, including counting how many are left, or counting on and counting back to find how many are left.

Taking away is not the only structure for subtraction; the 'comparison' structure is an important one for learners to experience. These are the *How many more?* and *What is the difference between … ?* type questions. A number line or track is very useful, particularly for counting on when finding the difference between two numbers.

Finding the difference

 Activity Book teaching notes

Page 18

This page has a focus on the comparison structure of subtraction. Learners need to use practical resources, such as counters or cubes, to make lines of different numbers. Model a problem for the learners, with two towers of 8 blue cubes and 5 red cubes. Ask them how they could work out many more blue cubes there are than red cubes. Demonstrate lining them up and looking at the difference between them. Count on from 5 up to 8 to show that the difference is 3. Repeat with other examples.

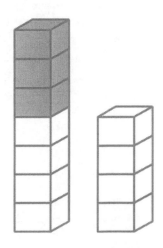

- In the one hand: Hold 4 counters in one hand and 7 counters in the other. Ask the learners to show how they would find out how many more there are in the right hand. As an alternative wording, ask learners to work out the difference between the number of counters in each hand. Look at the strategies they use, including lining the counters up to see the difference.

- Make them the same: Learners work in pairs, each taking a small number of counters, then comparing how many they have. One of them makes their pile the same as their partner, and says what they have done, for example: *You have three and I have five, so I have removed two of mine.* They take it in turns to change their counters.

- Maths washing line: Fasten numbers 0–10 to a washing line with pegs. Put rings round two of the pegs, such as 3 and 7. Discuss counting on from 3 to 7. Explain finding the difference. Repeat for different pairs of numbers.

- Floor number track 0–10: Use painted numbers on the playground or numbers written on card and placed in order on the ground. Ask the learners to hop forwards and backwards along the track, one number at a time. Discuss counting on, for example: *Stand on 3. Make 4 hops forward. Which number are you on now? Stand on 5. Hop on to 10. How many hops?*

Counting back; More counting back

 Activity Book teaching notes

Pages 19–20

- Counting back is the 'reduction' structure of subtraction, which is the inverse of addition when represented as an increase. For example, 8–3 with the idea of counting back involves starting at 8 and reducing it by 3, counting back to 5. A number line or track is an important visual representation to help learners understand this. Use PCM 6: Numbers lines 0–10 (on page 71 of this Teacher's Pack) for additional practice.

- Start by counting back together from 10 to 0 on the number track on page 19. Then choose different starting numbers and count back to zero. Follow this up by saying you just want to count back 3 (and then other numbers) from different starting points. Ask where the jumps will end.

- The language to use is: *Put the 8 in your head and then count back 3* before counting down *7, 6, 5* and showing this with your fingers.

- Floor number track 0–10: Use painted numbers on the playground or numbers written on card and placed in order on the ground. Learners hop forwards and backwards along the track, one number at a time. When hopping back, it is important that they do not turn round, but that they move backwards. Discuss counting back, for example: *Stand on 9. Make 3 hops back. Which number are you on now? Stand on 5. Hop back to 1. How many hops?*

- Down to zero: Each pair of learners starts with a tower of 10 cubes. The object of the activity is to get from 10 to 0. Turns are taken to remove 1 or 2 cubes at a time from the tower. After each turn, they write a number sentence to show what has been done for example: *10 count back 2 is 8. 8 count back 1 is 7.* Repeat for different quantities that can be taken away, such as 1 and 3, or 2 and 3.

- Spinning up and down: Provide a 1–3 spinner and a number track or line to 10. Use PCM 2: Spinners and PCM 7: Number track 1–10 (from pages 67 and 70 of this Teacher's Pack). Learners start at 5 on the track, use the spinner and count backwards or forwards by that amount. The object of the activity is to move up and down the track, visiting every number while counting on and back. Numbers that have been visited are crossed off on the track.

Hiding objects

 Activity Book teaching notes

Page 21

This page looks at the idea of hiding objects as missing number problems involving subtraction. Describe the example in the 'Learn' box at the top of page 21, basing it around a story of a bird visiting a nest and covering 3 eggs. Model it practically (using cubes or pretend eggs), showing that we can work out how many are hidden by looking at the number we can see. Show it as $5 - ? = 2$ and $5 - 2 = ?$ and talk about the ways of finding out the missing number. This is an opportunity to look at subtraction as the inverse of addition, if appropriate.

Activity ideas

- Hidden in the sand: In pairs, one learner hides a number of counters in the sand, while the other learners have their eyes closed. The learner who hid the counters tells the other how many counters he or she needs to find. The other learner searches until he or she finds the same number of counters that were hidden. As learners are searching, ask: *How many have you found? How many more do you need to find?*

- Hiding toys: Use up to 10 small toys and a yoghurt pot. Ask learners to count how many there are. Ask them to close their eyes and then cover some of the toys. Ask how many they can see and how many they think are hidden. Repeat for different numbers.

- Boxes: Learners draw boxes like these: ☐ − ☐ = ☐ They take a handful of cubes and write how many they have in the first box. They then take some away, and complete the boxes. This can be repeated for other handfuls of cubes. Tell them that 4 must be in the last box. They then explore possibilities for the first two boxes.

Success criteria

While completing the activities, assess and record learners who can:

- compare two cube towers and say how many more or less there are
- use a number track to take away by counting back
- work out the missing number of objects.

Unit 17 Measures and time

 Activity Book C, pages 22–25

 Story Book C

 Measuring mice, page 77
Clocks, page 78
My day, page 79

Learning objectives

- Use uniform non-standard units such as cubes to measure lengths.
- Use a balance to compare the mass of objects.
- Compare capacities of different containers and check by pouring.
- Recognise some hour times on analogue and digital clocks.

Resources

cubes, paperclips, lolly-sticks, counters, objects for comparing size, length and capacity, water tray, different transparent containers, balance scales, bottles, jugs, string, elastic bands, large analogue working clock, metal fasteners (split pins), bucket, sand timer, sand

Key words

compare, measure, length, height, order, capacity, full, empty, size, small, big, large, smaller, larger, bigger, long, longer, short, shorter, longest, shortest, tall, taller, tallest

Background information

The idea of conservation is a fundamental principle that applies to measurement. For example, a ball of modelling clay will weigh the same if it is made into a different shape, and a ribbon will still be the same length if it is moved. Practical measurement activities reinforce the idea of conservation and underpin the idea of transformation and equivalence. For example, when pouring a jug of water into a bottle, the volume of water stays the same.

Units of measure are introduced here, with the idea of moving on from direct comparisons of objects to the comparison of a single object to a number of equal units. Non-standard units are used to introduce the idea of a unit to measure with, such as cubes or pencils to measure the length of objects. This is an important step before introducing standard metric units, as it gives learners something familiar to measure with.

Measuring length

 Activity Book teaching notes

Pages 22–23

- Begin with the opening activity on page 22, asking the learners to decide how they can use a handful of cubes to measure the length of a pencil. Provide pencils and cubes and ask learners to work in pairs to measure. Discuss the accuracy: Did the learners click the cubes together or did they spread them out? They can then measure other items in the classroom using cubes.

- In the following activities, different units (such as cubes, counters and paperclips) are used to measure the length of objects. Talk about the similarities and differences between them and which are easy to measure with. They may not be exact units, so use the language: *The book is about 9 cubes long.*

Activity ideas

- Body measures: Each learner measures the distance round his or her head with string and cuts off this length. They find other parts of their body that measure about the same length as the string. They find items in the classroom that are about the same length. They may record their findings using simple drawings. There is no need for learners to compare any of their body measurements with those of others in the class.

- Spans and paces: Learners use paces to measure the width of the classroom. Record the results and talk about the fact that they are not the same for every learner. Ask why this is the case and talk about using units that are all the same length. Repeat this for hand-spans for the length of a table.

- Measuring mouse: Make a 'measuring mouse' for each learner using PCM 12: Measuring mice (on page 77 of this Teacher's Pack). Ask learners to measure different objects using the mouse and compare lengths by the number of 'mice' long they are.

Weight and capacity

 Activity Book teaching notes 61

Page 24

- There are two main aspects of measure to focus on in this activity page and it is probably easier to manage if you teach each separately. Young learners' experience of weighing should begin with balance-type weighing equipment. Concentrate on comparisons: *Which are heavier? Which are lighter?* and then focus on weighing items so they are the same as other items. So, for example, a book weighs the same as 10 cubes. The important phrase is 'weighs the same as' so that equivalence is being experienced with non-standard units. This will lead on to using standard units, measuring items that 'weigh the same as' a mass of 200 grams or 3 kilograms, for example.

- Learners can confuse weight with size, so include items to compare that are similar sizes but different weights. Concentrate on comparisons. Ask: *Which are heavier? Which are lighter?*

- Conservation of capacity with different-shaped containers is an important concept to master. Use lots of different containers in the water tray before doing the activity: fill one and pour into different containers to compare. Learners can then see which containers are the largest and smallest.

Activity ideas

- Elastic-band lines: Place elastic bands on two different containers. Pour water into the first container to the level of one band, and then pour the water into the other container. Ask: *Is the water above or below the line?* Discuss which holds the most/least.

- Pouring out: Learners place elastic bands on the transparent containers, and then pour water to the level of the band as accurately as possible. They then fill each container to the top, and then pour the water out. They stop pouring when they think the water has reached the level of the elastic band.

- Balances: Use balances and a variety of different materials. Discuss balancing and making each side 'the same'. Talk about 'heavier', 'lighter' and 'the same weight'.

Time

 Activity Book teaching notes

Page 25

- Learning to tell the time can be tricky for some learners. This is particularly true with an analogue clock due to the linear scale in a circle, and the numbers representing both the hours and the minutes past, depending on which hand is pointing to it.

- For the activities on this page, give learners small clocks to use at the same time. Make clocks for the learners using a split pin and the template on PCM 13: Clocks (on page 78 of this Teacher's Pack) on card. Focus on the short hour hand pointing to the different numbers to show each hour passing and make sure they understand that the minute hand is 'stuck' at 12, which shows 'o'clock' or the exact hourly time. This will be fairly abstract for learners, but it will make sense for them when they get on to 'minutes past the hour' in their next stage.

- An hour is quite a long time for learners to have a real understanding of. Minutes and seconds are easier to practically experience, so use activities alongside the language of time to build up their understanding of these shorter periods of time.

Activity ideas

- 'What's the time?': An analogue clock should be available to look at during the day. Draw the learners' attention to the time at significant points in the day, such as lunchtime. Emphasise that the short hand points to the hour – concentrate on this rather than showing the minutes past the hour.

- Ordering time: Find pictures of events during the day and make them into a set of 5 or 6 cards, or use the cards on PCM 14: My day (on page 79 of this Teacher's Pack). Ask learners to match these with a clock to show the times to the hour that the pictures could represent. They do not have to be accurate with this, just assign times to the cards. They can then put the cards in order of time during the day.

- Sand timer: Ask learners to guess how many times they can fill a bucket with sand and empty it again in one minute. Set them a minute sand timer and check their guess, asking another learner to count how many. Ask them to use the minute sand timer to count how many times they can carry out a different task.

Success criteria

While completing the activities, assess and record learners who can:
- measure the length of a table using cubes
- use a balance to find out which objects are heavier than others
- compare the capacity of containers and say which is the most full
- read some times on a clock.

Unit 18 Shapes and patterns

 Activity Book C, pages 26–29

 Story Book C

 Shapes, page 80

Learning objectives

- Name flat shapes, describing some generalised properties of each shape.
- Identify shapes in different positions and orientations.
- Make repeating patterns with shapes.
- Recognise the face shapes of solid shapes.
- Name some solid shapes and describe properties.
- Recognise simple shapes and objects that show reflection and symmetry.

Resources

2-D shape tiles, 3-D solid shapes, patterned fabrics and wallpaper, beads, laces, cut-out shapes, hoops, paint, construction kits, sticky-backed paper, glue-sticks, butterfly outlines on sugar paper

Key words

flat shapes, solid shapes, circle, triangle, square, rectangle, cube, cuboid, cone, cylinder, face

Background information

- Your learners will recognise and name certain 2-D and 3-D shapes and will be beginning to reason about their properties. Concrete manipulatives are important and learners need experience with these and also how the shapes are represented on paper. The orientation of a shape does not alter its properties – this needs emphasising.

- Symmetry is a new idea when looking at properties of shapes and objects. A shape that has reflective symmetry remains unchanged if its side or sides are reflected through a line drawn from the middle of the shape. We are not introducing lines of symmetry yet, but the emphasis is on two halves of a shape looking the same.

Flat shapes

 Activity Book teaching notes

Pages 26–27

- Page 26 is mainly a 'talk-about' explore page, with a picture full of shapes to find and discuss. Talk about similarities and differences between the shapes and where you see them around the classroom and school. Reinforce counting carefully when finding the shapes in the picture.

- The patterns on page 27 use the two attributes of colour and shape. Learners could use beads or shape tiles to make patterns and then copy them to show each of the attributes.

Activity ideas

- I spy shapes: Play 'I spy shapes' with your class. Start with: *I spy with my little eye, an object in the class room that is the shape of a circle.* Learners put up their hand when they see a circle in the classroom, such as a clock, a picture of a sun or a wheel. Repeat with other shapes.

- Cut-out shapes: Give learners a variety of cut-out shapes (either on sticky-backed paper or with glue-sticks) and a large piece of paper and ask them to make a picture from the shapes. They might make a 'sun' using a circle, a 'window' with a square or a 'door' with a rectangle.

- Limit the tiles: Give learners a limited number of shape tiles, for example just triangles and squares. Ask them to make pictures just using these shapes. Talk about the objects they have made and ask: *How are they the same or different?*

- Shape search: Give a sheet of paper with 4 large shapes drawn on – triangle, square, circle and rectangle. Use PCM 15: Shapes (on page 80 of this Teacher's Pack). Ask learners to walk around the classroom (or outdoors among play equipment) and tick inside each shape when they see a shape like this around them. Talk about which shape has the most ticks and which shapes were difficult to find.

Solid shapes

 Activity Book teaching notes

Page 28

- Give learners 3-D shapes to handle and compare them to the pictures of the shapes on page 28. Ask the learners if they know the names of the shapes. Repeat the names so they can learn them, and so they are pronounced correctly.

- Talk about the similarities and differences between the shapes and focus on the faces of each. Explain that a sphere has one curved face, a cone has a curved face and a flat face, and a cylinder has a curved face and two flat faces. Ask the learners to name the shapes of each face.

Activity ideas

- Show me: Learners each have a cube, cuboid, cylinder and sphere. Ask them: *Show me a shape that … will roll…, has a flat face, … is round, … that can be stacked.* Then ask: *Show me a shape that … cannot be stacked, … has no flat faces, … has no curved faces, has 6 faces.* Ask why they give some of their answers to check their understanding of the properties of shapes.

- Rules for shapes: Learners sort a mixed set of flat and solid shapes. Use a hoop on a table and ask them to put shapes inside the hoop that all have the same 'rule' for sorting that they can choose. You can guess their rule or ask them to explain their rule. Continue by asking them to sort the shapes again, using a different rule.

- Pass it on: Learners sit in a large circle. Give a solid shape to be 'passed on' round the circle. As the shape passes from learner to learner, a different fact should be given about the shape. Encourage a co-operative approach where everyone is helping to get the shape as far round the circle as possible.

Symmetry

Activity Book teaching notes

Page 29

If a shape or object is symmetrical, we are, in this case, looking at the property of reflective symmetry. Look at the objects in the 'Learn' section at the top of the page and ask the learners what they notice and what makes the shapes symmetrical or not symmetrical. Talk about two sides being the same and make sure they can see this. Draw a line down the centre of a symmetrical shape drawn on the board as that may help.

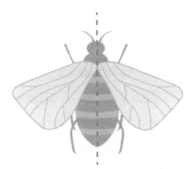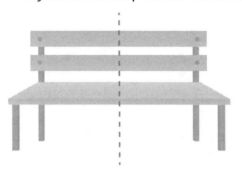

Activity ideas

- Butterfly paint drops: Draw some butterfly outline shapes on sugar paper and ask the learners to put drops of different colour paints on one wing. Fold the wings over on top of each other before the paint dries and check to see if the wings are symmetrical.

- Symmetrical models: Ask learners to use construction kits such as interlocking bricks, 3-D interlocking shapes, geometric shapes or interlocking cubes to make models that have symmetry. Ask them to describe their models and why they think they are symmetrical.

- Same and different: Show two flat shapes that are the same, but look different, such as two triangle shape tiles. Ask: *Are these the same shape? How are they different?* Show different shapes and let learners say how they are different and how they are the same. Make sure these include symmetrical and asymmetrical triangles.

- Halves: Learners work in pairs. Each makes identical halves that are placed together to make a 'whole'. They can use interlocking cubes, interlocking bricks, building bricks and geometric shapes.

Success criteria

While completing the activities, assess and record learners who can:

- describe flat shapes and know what makes them the same or different from others
- recognise shapes when they are in different positions
- make repeating patterns with shapes
- describe the faces of different solid shapes
- know the names of some solid shapes
- show a symmetrical shape.

Assessment

- Ask learners to complete the *What can you remember?* activities on pages 30–31 of Activity Book C.
- Ask learners to self-check their understanding of key objectives covered this term, using the Self-assessment chart on page 32 of Activity Book C.

Number cards 0–10

0	1	2
3	4	5
6	7	8
9	10	

Hodder Cambridge Primary Maths Foundation Stage Teacher's Pack © Hodder & Stoughton Ltd 2018

Spinners

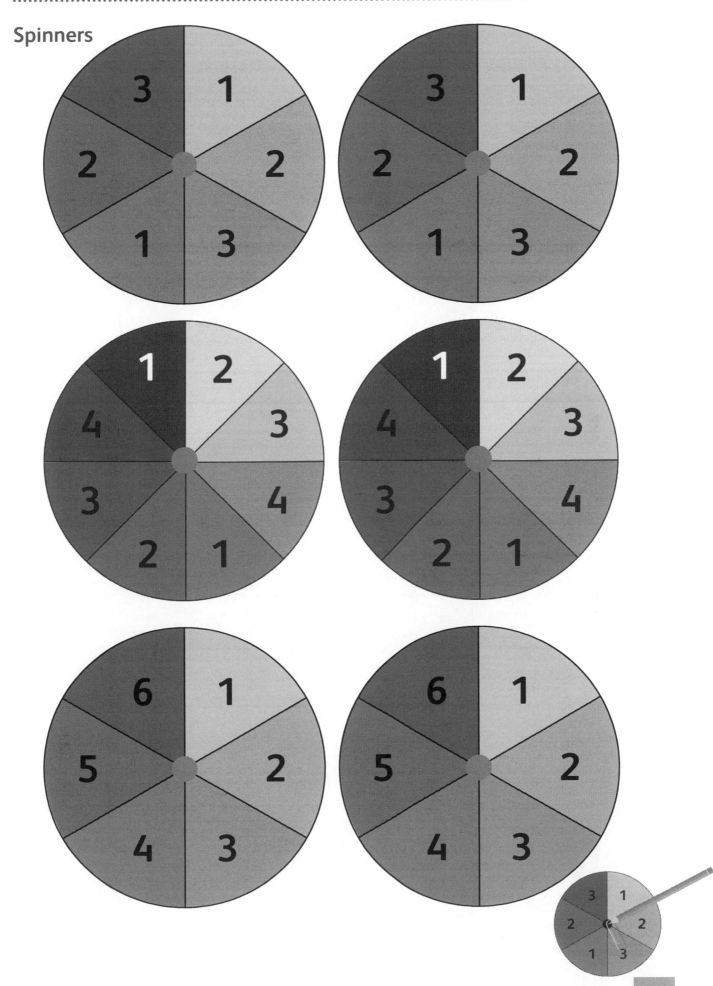

Dominoes

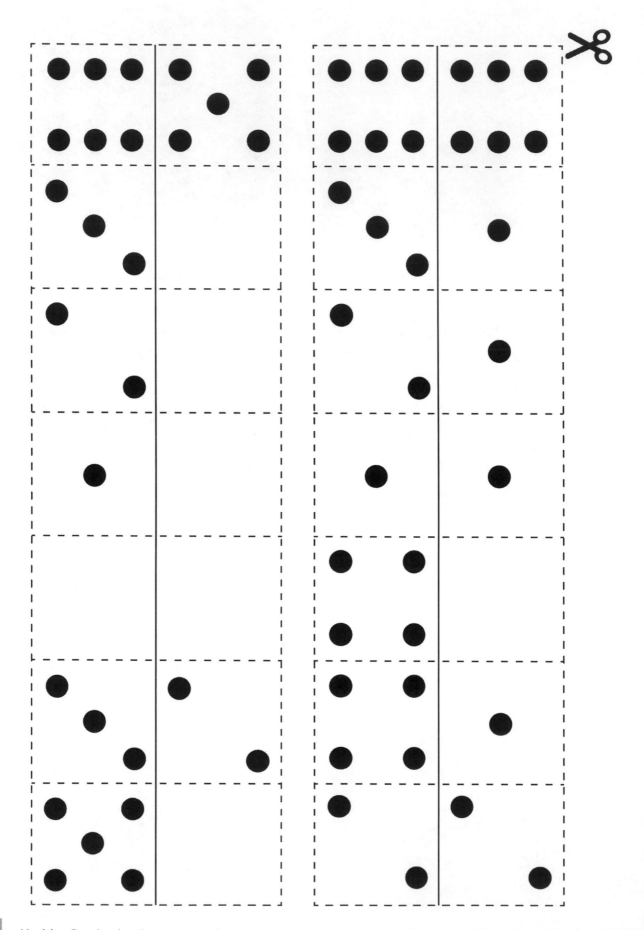

Hodder Cambridge Primary Maths Foundation Stage Teacher's Pack © Hodder & Stoughton Ltd 2018

Dominoes

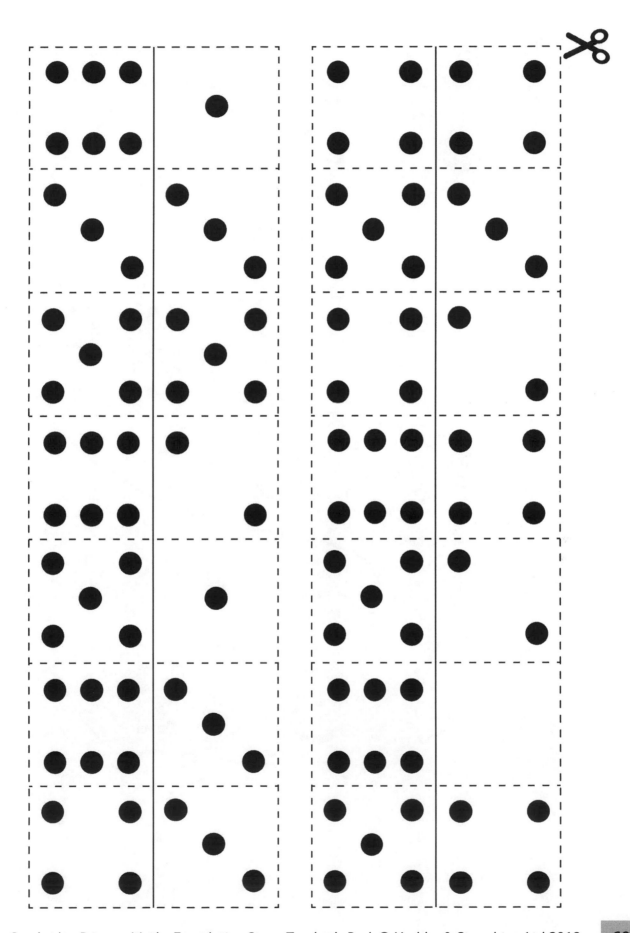

Number track 1–10

Number lines 0–10

Number track 1–20

Number lines 0–20

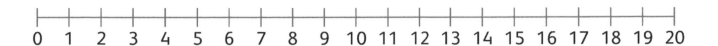

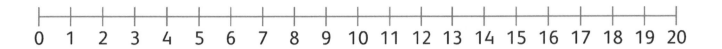

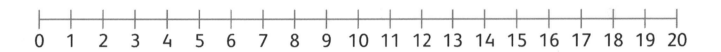

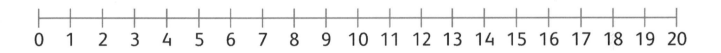

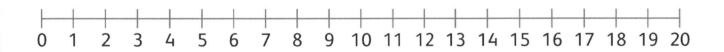

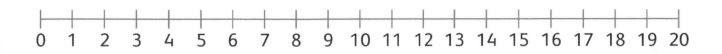

Five frames

Hodder Cambridge Primary Maths Foundation Stage Teacher's Pack © Hodder & Stoughton Ltd 2018

Ten frames

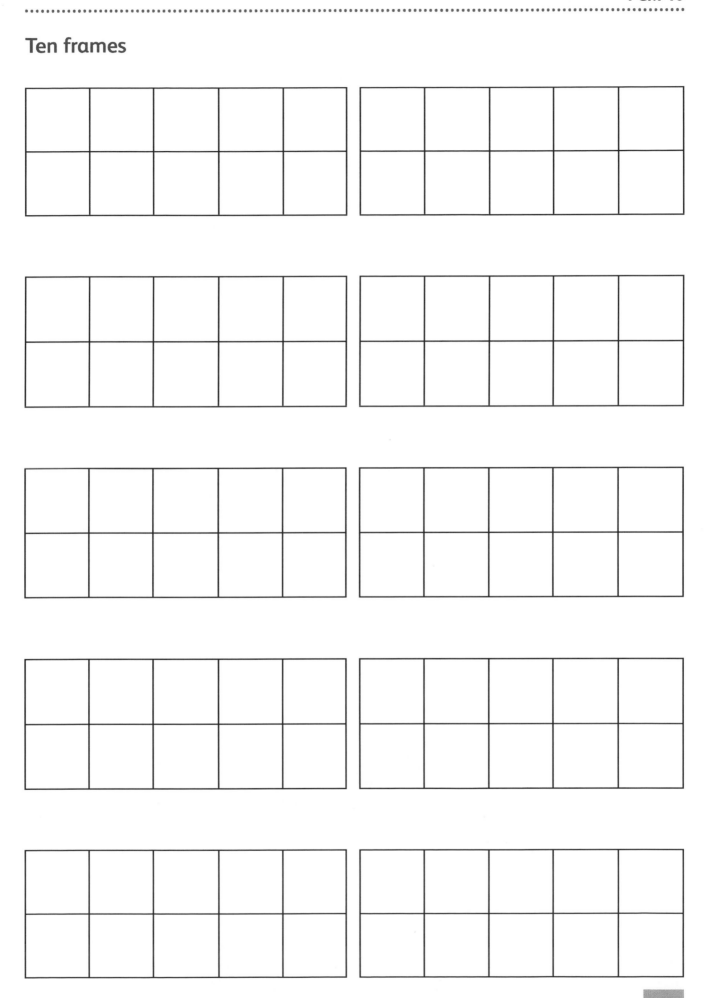

Partition circles

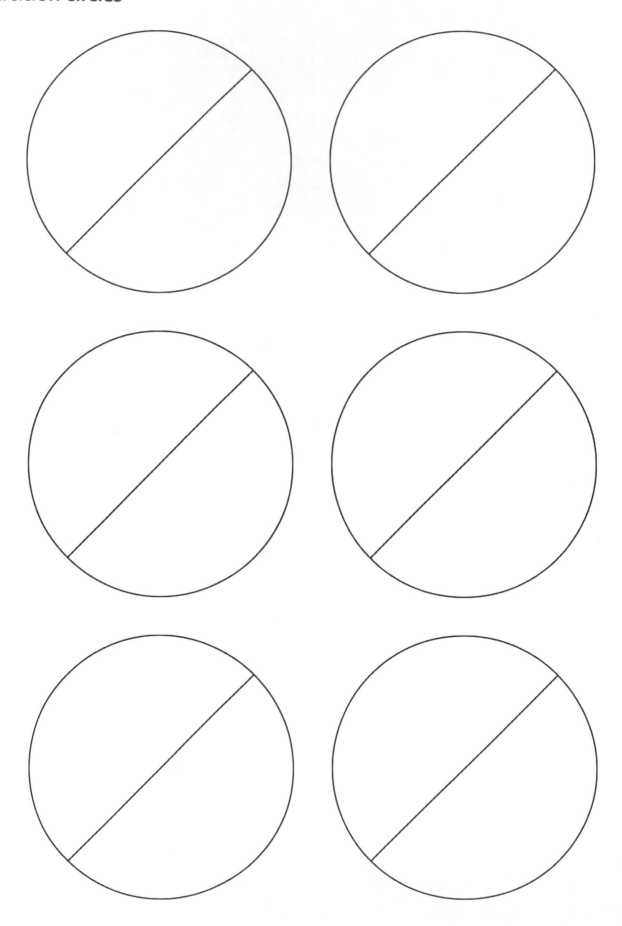

Hodder Cambridge Primary Maths Foundation Stage Teacher's Pack © Hodder & Stoughton Ltd 2018

Measuring mice

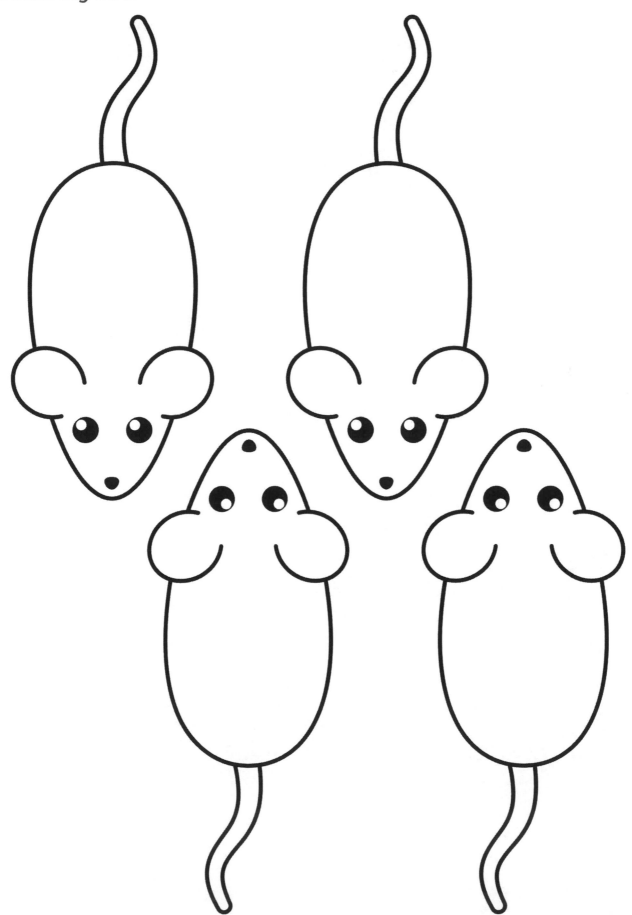

Clocks

My day

Shapes

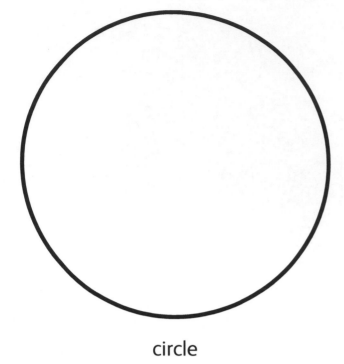

circle

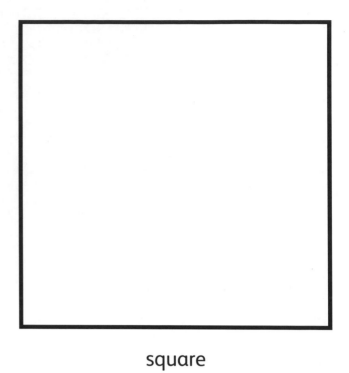

square

rectangle

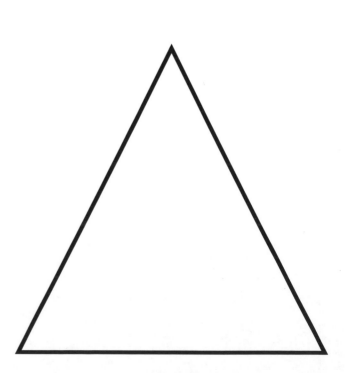

triangle